Elastane in Sports and Medical Textiles

Elastic garments have tremendous scope in the field of tight-fit sportswear and healthcare applications. Research towards improvement in the elasticity of fibre, yarn, and fabrics and development in testing methods for elastic garments is the current requirement for the industrial product development. This book covers elastane fibres, elastane yarn and fabric production methods, new attempts in yarn production, commercial ways of fabric-manufacturing techniques and the fabric properties, new testing methods to test the elastic products, and applications of elastic garments in sports and healthcare.

Features:

- Provides comprehensive review, process, and application of elastane fibres.
- Covers detailed information about manufacturing and testing methods of elastane fabrics.
- Reviews technical aspects of elastane in sportswear and healthcare.
- Discusses evaluation process for the elastane fabric performance.
- Defines production methods of woven and knitted fabrics using elastane.

This book is aimed at students and researchers in textile engineering/technology, textile design, human ecology and comfort, material engineering, sports sciences, medical science, and healthcare engineering.

Elastane in Sports and Medical Textiles

R. Rathinamoorthy and
M. Senthilkumar

CRC Press
Taylor & Francis Group
Boca Raton London New York

CRC Press is an imprint of the
Taylor & Francis Group, an **informa** business

CRC Press
Boca Raton and London

First edition published 2023
by CRC Press
6000 Broken Sound Parkway NW, Suite 300, Boca Raton, FL 33487-2742

and by CRC Press
4 Park Square, Milton Park, Abingdon, Oxon, OX14 4RN

CRC Press is an imprint of Taylor & Francis Group, LLC

© 2023 Taylor & Francis Group, LLC

ISBN: 9781498779548 (hbk)
ISBN: 9781032409252 (pbk)
ISBN: 9780429094804 (ebk)

DOI: 10.1201/9780429094804

Typeset in Times
by codeMantra

Contents

About the Authors

Dr. R. Rathinamoorthy has been working as an associate professor in the Department of Fashion Technology, PSG College of Technology, Coimbatore, India, since 2009. He completed his PhD in Medical Textiles in 2016. Recently, he received the Young Achiever Award 2019 from the Institute of Engineers India (IEI), Coimbatore chapter. Also, in 2017, he received a national-level Young Engineer Award 2016–17 in the textile engineering domain from IEI, Kolkata, West Bengal, India. He has a Google H index of 19 and a Scopus H index of 13 with more than 1020 citations. He has published 20 national and 73 international research articles in various refereed and non-refereed journals. He has authored eight technical books in the area of apparel and fashion technology and has 28 book chapters with various international publishers to his credit. His research interests are sustainable materials and environmental sustainability as related to the textile industry. He is currently working on biomaterials for textiles uses and microfibre pollution generated from textiles.

Dr. M. Senthilkumar has been working in the Department of Textile Technology, PSG Polytechnic college, Coimbatore since 2006. He completed his PhD in 2012. He has published more than 50 articles in international and national reputed journals. He received Young Engineer Award 2014 for the textile engineering discipline instituted by the Institution of Engineers (India). He is actively involved in academic as well as research. He is presently working in the field of active sportswear.

Foreword

Potential applications of elastane and their textile products in the field of sport and medical fields are growing day by day. Elastane medical products like compression bandages and stockings are used for scar management, controlling blood pressure, and reducing muscle strain and sprains. Similarly, elastane sports products like compression garments enhance the sport performance of the sportsperson. Dissemination of the knowledge and the awareness on elastane yarn and its fabric production for specific end-use is essential for the people who are manufacturing these products and who are involved in new product developments. The primary objective of publishing this consolidation work is to make it useful to the student community, academicians, industrialists, and technocrats who wish to understand elastane and its applications in different fields.

This book deals with the chemical structure and production methods of elastane fibre, physical and chemical properties of the elastane fibre, and their potential applications. Further, fabric properties like physical, mechanical, moisture, and thermal and elastane-specific properties like stretch and recovery, compression, and handle have been discussed. Effects of elastane content on the body-fit and comfort characteristics of elastane fabrics are also elaborated. The application of elastane-incorporated fabric in compression stockings, bandages, and body-shaping garments and other applications in compression therapy are enlightened. The application of compression garments in the area of sports and aesthetic wear is also discussed. Selection of fabric production methods for various medical applications; mechanisms and functions of compression stockings, bandages, and orthopaedic supports; and usage of compression garments in active sports are highly informative.

This book is useful for all stakeholders as it leads to better understanding of elastane fibre, yarn, and fabric and their functions during various applications. In my view, this book is extremely valuable and timely addition to the technical literature.

Dr. V. K. Kothari
Retd. Professor, IIT Delhi

Preface

Elastane in Sports and Medical Textiles covers the technologies and methods used in the manufacturing of stretch fabric and details their potential market applications. The stretch fabric has become one of the most used commercial fabrics due to the prevailing fashion trend. However, the knowledge of the elastane fibre, production methods, and their impact on fabric properties are not much discussed. Though several research works were performed on the applications of elastane fabric, those researches were more specific to the particular fabrics, structure, and method of elastane insertion. Hence, it's been our primary objective to amass the fundamental and recent research advancements in elastane fabrics. This consolidation will serve as a textbook for the next generation of students, academicians, and technocrats who wish to understand the fundamentals or wish to perform their research in this field.

This book provides insight to the readers about the elastane fibre, yarn, and fabric production with the details of additional setups required to insert the elastane yarn at every stage of production. The book also covers the effect of various elastane insertion parameters and yarn properties on the properties of elastane fabric or stretch fabric. The book details the fabric development and properties of both knitted and woven fabrics. The latter part of the book outlines the application of elastane fabric in medical, sports, and other fashion applications. This book is a first of its kind in the textile and fashion discipline that entirely covers the fundamentals of elastane yarn production, fabric production, knit and woven fabric production, and their applications.

Chapter 1 is an introduction chapter, where the fundamental details of the elastane fibre are summarized. The first part of the chapter outlines the chemical structure and production methods of elastane fibre. The last section of the chapter details the physical and chemical properties of the elastane fibre along with their potential application. Chapter 2 outlines the elastane yarn and fabric production methods. In this chapter, different methods used in elastane yarn production are described in the first part. The knitted and woven fabrics production methods with elastane yarn are detailed in the latter part of the chapter.

Chapter 3 delineates the elastane-incorporated woven and knitted fabric properties. Fabric basic properties like physical, chemical, mechanical, moisture, and thermal are detailed with the latest research results. The elastane-specific properties like stretch and recovery, compression, and handle are also elucidated in this chapter.

Chapter 4 consolidates the effect of elastane content on the fit and comfort characteristics of both knitted and woven fabrics.

Application parts of the elastane-incorporated stretch fabric are discussed in Chapters 5 and 6. In Chapter 5, the application related to the medical industry is detailed. The chapter discusses the application of elastane-incorporated fabric in compression garments, bandages, and body-shaping garments, and other applications in compression therapy.

Chapter 6 reports the application in the area of sports and aesthetic wear. In this chapter, the use of compression garments, the effect of fit, and aesthetic properties along with stretch, recovery, and comfort properties are addressed.

Through these six chapters, this book consolidates all relevant information on elastane fabric from fibre to production and applications. It is the hope and confidence of the authors that this book will provide complete know-how about elastane-incorporated textiles.

Dr. R. Rathinamoorthy

Dr. M. Senthilkumar

1 Elastane fibres – production and properties

1.1 INTRODUCTION

Elastomeric fibres were developed in the time of World War II to replace the requirement of natural rubber. In 1952, the production process on elastomeric fibre was patented in Germany [1]. As the war required existing rubber for building equipment, the price and availability of the rubber greatly varied in time. Hence, scientists took an attempt to develop a synthetic alternative to natural rubber at the time of World War II. The first polyurethane elastomers were produced from millable gums. At that time, Du Pont produced the first Nylon polymer; however, they were stiff and rigid. On further research, Du Pont developed a unique form of polyurethanes that could be made into fine threads [1]. In 1950, Du Pont made the first breakthrough by introducing an intermediate substance in the Dacron polyester and made a stretchy fibre. After a decade of research, chemist Joseph C Shivers improvised the fibre and released it to market in 1958. To brand the fibre, Du Pont selected the name Lycra as a trade name, whereas it was originally known as Fiber K [2, 3]. The elastomeric fibres are also called in the name of Spandex. It is not a brand name but an anagram of the word expands as reported by Kadolph (2006) [4]. The word Spandex is commonly used in North America and Elastane is common in Europe [5, 6]. Later in the 1980s, due to the extensive campaign and increased fitness trend among the public, the sales of spandex exploded and Du Pont was not able to produce the required quantity that the market demanded [1].

Elastomeric fibres are synthetic segmented copolymers, developed from polyurethane copolymers. These fibres can stretch up to 200% upon load application and can be returned to their original shape without residual changes [7]. Elastane, one of the most important thermo-plastic elastomeric fibres being commercially produced worldwide, is made with long-chain synthetic polymers comprising mostly segmented polyurethanes [8]. In other terms, the elastane yarn consists of a polyglycol long-chain unit joint with a short di-isocyanate compound and the fibre contains at least 85% polyurethane [9]. By nature, the chemical structure of the elastane fibre contains alternative hard and soft segments linked through urethane bonds. The elastic nature of the fibre is attributed to the soft structure of the polyurethane block copolymer, and the hard segment develops molecular interaction to provide the required stability and strength to the fibre [10]. The fibre is elastomeric, which

DOI: 10.1201/9780429094804-1

means it can be stretched to a certain degree and it recoils when it is released. These fibres are superior to rubber because they are stronger, lighter, and more versatile [11]. DuPont has been the world's leading manufacturer of elastane fibre in the commercial name of "Lycra" since 1962 [12, 13]. The main advantage of the elastane fabric is that these fabrics fit on the body in perfect shape and retain it throughout the wear [14].

In recent times, the usage of elastane has increased due to an increase in awareness of garment fit and healthcare. Researchers have projected a spandex market growth rate (CAGR) of 2.2% from 2021 to 2027. The market value of the spandex was around 7.39 billion USD in 2019 [15]. The elastane fibre market is majorly driven by the clothing sector's huge contribution (74%) followed by the medical and other applications. Growing awareness about the healthy lifestyle which increased the activities like sports, exercise, and leisure activities was noted as the major reason for increased consumption by clothing sector. In 2019, the market value of the US elastane (spandex) was 1.5 billion USD and it is projected to grow in a CAGR of 2.7% from 2021 to 2027. The Asia-Pacific region contributed with maximum market share with respect to the elastane fibre [16]. This chapter outlines the various production methods, physical and chemical characteristics of elastane fibre, and their yarns in detail.

1.2 RAW MATERIALS AND ELASTANE PRODUCTION

For the production of elastane fibres, two different types of prepolymers are used to react and form the backbone of the elastomeric fibres, namely, macroglycol (the flexible component) and di-isocyanate (the hard component). Out of these, the macroglycols are generally of common polymers like polyester, polyether, polycarbonate, and polycaprolactone. The structure of these macroglycols consists of hydroxyl groups on both ends of the structure (–OH). These groups readily react with hard segments to form the chain, and also these groups are responsible for stretching characteristics of elastomeric fibres. The di-isocyanate groups are the responsible polymers for the hard segment. They are made of isocyanate (–NCO) groups on both sides of the structure, and it provides the required strength to the polymer chain. Other than these fundamental ingredients, other materials like catalyst, stabilizers, colouring agents, and finishers are also included in the structure of the polymer. Catalysts are used to initiate the reaction between two prepolymers during the production of elastomeric fibres. Di-azobicyclo octane and amines are the common catalysts used to initiate the reaction and also to control the molecular weight of the fibre, respectively. Stabilizers are included in the fibre structure to protect the fibre from external damages like heat, UV light, and other atmospheric contaminants. As the elastomeric fibres commonly used in swimwears, stabilizers are usually used along with the polymers during the production process to withstand the effect of mildew, and to prevent decolouration [17].

The elastane yarn production method consists of three different stages as reported by Gorden Cook [18]. The first step consists of the production of a low-molecular-weight polymer as a prepolymer for the final production. This is mainly performed to create a segment in the polymer chain. In this process, the formation of the amorphous region helps in creating flexibility and elongation characteristics due to the unfolding of molecules. The soft segment structures in the polymer chain mainly influence the fibre properties like melting point, flexibility, and chemical stability [18]. In this process, hydroxyl groups in the macroglycols interact and establish bonds with isocyanates. In this reaction, through step-growth polymerization, each molecule gets added to the end of another polymer molecule and forms the long-chain backbone for the elastomeric fibre [17]. In the second step, the developed low molecular weight prepolymer is allowed to react with the di-isocyanate for the chemical reaction. In this step, the segmented polyurethane is converted into a soft segment polymer. During this process, the hydroxyl groups in the soft segmented polymers react with the isocyanates and form urethane groups. Due to their higher reactivity, aromatic di-isocyanates are most commonly used in the development of elastane (spandex) polymers. At the end of the process, the reaction between the di-isocyanate and prepolymer is ended to produce a segmented polyurethane polymer. In this process, the polymer is exposed to glycol or diamine and this will result in a hard segment with the formation of urea or urethane groups [18]. The formation and chemical structure of the elastane (spandex) is provided in Figure 1.1.

The common methods used in the production of elastane fibre are dry-spinning, reaction spinning, and wet/melt spinning.

The dry and wet spinning methods are used for the production of linear polymer that can be melted and extruded. Whereas, the insoluble, hard, and non-melting types of elastane are produced through reaction spinning. Out of all these methods, the majority of the manufacturers use the dry spinning method to produce the elastane fibre. In this process, polyurethane is dissolved into a solvent and then

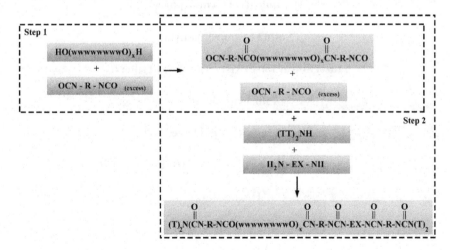

FIGURE 1.1 Elastane fibre formation mechanism.

extruded into a controlled atmosphere to remove the solvent [18]. In the case of reaction spinning, the isocyanate-terminated pre-polymers are extruded in a chain extender bath and it helps in developing the end product [19]. In the dry spinning process, after the addition of diamine, a solvent is added to dilute the solution. This diluted solution is pumped through a spinneret to form filament strands. By exposing the strands to solvent gas like nitrogen, the liquid polymer is hardened in the dry spinning. In order to develop a required thickness, several single filaments are combined together to form a multifilament. A slight twist inserted during the drying aids the formation of twist and grouping the multiple filaments. After the application of necessary finishing agents, the filaments are wound on a spool or cheese based on the requirements [20]. The dry spinning process sequence is provided in Figure 1.2.

After the manufacturing process, the elastane fibres can be scoured, bleached, dyed, and finished as per requirement. Generally, scouring of elastane yarn is performed using the detergent to remove the traces of solvents that are used in the dry or wet spinning process. The scoured yarns were then bleached to improve their whiteness index using peroxide bleaches. As the other bleaches like sodium and

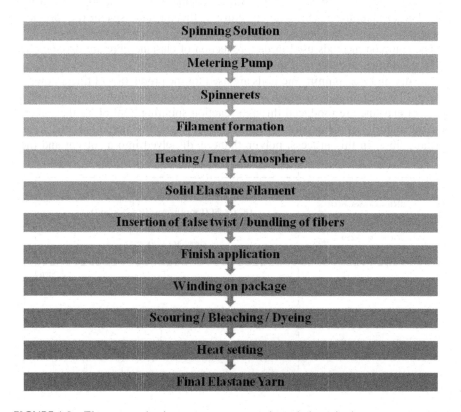

FIGURE 1.2 Elastane production process sequence through dry spinning.

hypochlorites result in discolouration, peroxides are preferred much in this case. Acid dyes are the commonly used dyestuff for elastane as they consist of basic reactive sites. Various kinds of resin finishes also can be applied to the elastane yarn to improve its handle and resistance against contaminants. If necessary, the elastane fibres can also be passed through a heat setting process like other thermoplastics, to improve their dimensional stability [18].

1.3 CHEMICAL STRUCTURE

Elastane fibres are generally blocked copolymers. The arrangement of low melting (amorphous) and high melting (hard) segments in a single polymer chain provides the expected extension in an elastane yarn. As discussed earlier, the commercial elastomeric fibres consist of aromatic polyureas as hard segments and the soft segments are mainly made of urethane linkage. The properties like stability against the temperature, modulus, and property of elongation are influenced by the physical size and shape of the hard segment molecules [21]. However, higher extensibility and recovery of the fibre are largely affected by the soft segment properties like low interchain bonding, low melting point, and low internal viscosity [19]. During the spinning process, the molecules are arranged in a randomly disordered fashion as coils. This helps the elastane to unfold when the stretching force is applied and results in extension. These coiled soft segments and hard segments are connected by strong intermolecular links by hydrogen or van der Waal's forces. The connection acts as a secure point and avoids the sliding of molecules during the stretching process. Hence, after the release of stretch, the extended molecule return to its folded or coiled state and reaches its original length [18]. The molecular arrangement of elastane fibre is provided in Figure 1.3.

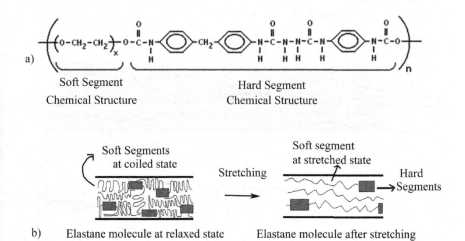

FIGURE 1.3 Elastane fibre molecular arrangement with soft and hard segments.

1.4   PROPERTIES OF THE ELASTANE FIBRES

1.4.1   PHYSICAL PROPERTIES

The Elastane fibres are produced from spinneret as monofilaments and it forms the shape of the spinneret. In general, they are circular. Elastomeric fibres are block copolymers that provide the manufacturers with a higher level of control over the structure. For instance, the following can be controlled during the production,

    i. the type and length of the soft segments
    ii. the type and length of the hard segments and
    iii. the type and degree of cross-linking can.

The correct use of cross-linking controls the properties of the polymer to a greater extend. The cross-link is the process that creates a bond between adjacent polymer chains. These bindings are extremely important in the physical properties of the polymers. Very few cross-link points in a polymer will reduce the solubility of the polymer and will develop polymer with higher swelling capacity like gel as shown in Figure 1.4(b). Whereas, *highly cross-linked* polymers develop hard, infusible, and insoluble products having a three-dimensional network as shown in Figure 1.4(c) [22].

The elasticity, flexibility, and extensibility of the elastane fibres are mainly due to the presence of lower molecular polyester or polyether content in their structure. The hard segment in the structure offers modulus and toughness to the polymer structure. These fibres are nearly transparent in colour, dull in lustre, and can be dyed as required. One of the main advantages of the elastane fibre is that the shades of these fibres not necessarily to be matched with other fibres. As the elastane fibres are hidden inside the structure, the transparent property will help to attain the required shades. However, the elastane materials can be dyed with disperse and acid dyes including

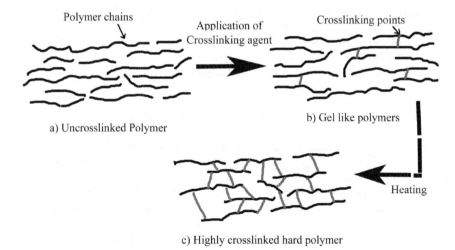

c) Highly crosslinked hard polymer

FIGURE 1.4   Role of crosslinking agent on the production of elastane material.

TABLE 1.1

Physical properties of elastane fibres

Number of filaments in the yarn	Normally 5–50
Linear density of filaments tex (g/km)	0.1–3
Density (g/cc)	1.2–1.25
Moisture regain (%)	0.3–1.2
Tensile strength (kg/cm²)	616–994
Breaking Tenacity (grams/denier)	0.55–1.0 (4.9–4.8 cN/tex)
Elasticity (%)	400–800
Elastic recovery	Permanent set noted after repeated stretching. Recovery increased above 300% stretch
Modulus of elasticity	Lower modulus of elasticity (1/1000 of a conventional hard fibre like cotton)
Heat	Degrades over 150⁰C
Flammability	Burns quickly
Electrical conductivity	Low and not damaged for any conductive radiation
Melting point	Starts at 175⁰C, melts at 250⁰C
Sunlight	Highly resistant. Prolonged exposure causes decolouration
Age	Deterioration noted with ageing
Dyeing ability	Good
Abrasion resistance	Very good

the chromed and metallized dyes [23]. The other physical properties of the elastane fibre are provided in Table 1.1 as reported by other researchers [18, 24].

1.4.2 CHEMICAL PROPERTIES

Elastane fibres are affected by an oxidative attack by atmospheric contamination. Hence, during the manufacturing process itself, manufacturers use to include several resistive chemicals into the spinning solution. These atmospheric contaminants degrade the polymer by a free radical chain reaction. Due to the presence of double bond, natural rubber provides strong resistance against these contaminants. The use of polyether in the soft segments of the spandex increases the easiness of the attack. Hence, the use of an effective stabilizer is necessary to have a resistive effect on the elastane material. Chlorine bleaches are the most highly affecting chemical in the case of synthetic elastane fibres as they directly attack the hard segments. Hence, for bleaching of elastane, the use of peroxide bleaches is highly recommended. The presence of soft segments in these polyester fibres was noted as one of the reasons for its susceptibility to mildew attacks in the swimwear. Hence, most often antimildew agents will be added to the commercial products. The elastomeric fibres also swell into some organic solvents like hexane (rubbers) and chlorinated solvents (spandex). The effects may be temporary as the fibre will be back to normal once the solvents are evaporated. However, this process may remove some of the stabilizers in the

TABLE 1.2

Chemical properties of elastane fibre

Properties	Elastane fibre
Reaction with acid	Good resistance for mild acids
Reaction with alkalis	Good resistance to most of the alkalis.
Reaction with organic solvents	Resistance to dry cleaning solvents. Unsaturated hydrocarbons may damage on prolonged exposure.
Fastness	Good washing fastness and fairly good light fastness with acid dyes
Reaction with bleaches	Degraded by sodium hypo-chlorite & peroxide is preferable
Dyeing	Affinity towards acid and disperse dyes
Laundering	Washing at 60°C and tumble drying at 80°C
Stability to UV light, cosmetics and ozone	Good resistance
Stability to NO_x and active Cl	Fair and turns yellow by exposure
Insects	Completely resistant
Microorganism and mildews	Completely resistant
Common chemicals	Good resistance

structure and spoil its quality [24]. Change of colour to yellowness is one of the common issues addressed with elastane when it is exposed to sunlight or UV radiation. Researchers showed much interest in controlling this degradation by incorporating pigments and additives to maintain the white colour and restrict discolouration. Exposure of the fibres to atmospheric contaminants or oxidation agents, heat, and light reacts with the aromatic segments of the elastane fibre and results in discolouration. Researchers recommended the use of phenolic and amine-type stabilizers during the manufacturing process as an antioxidant to avoid discolouration. The other chemical characteristics of the elastane fibres are provided in Table 1.2, as reported by Gorden Cook [18].

1.5 APPLICATIONS OF ELASTANE FIBRE

The elastane fibres are used in various applications, and mainly on clothing. The elastane fibres are commonly used in athletic wear as it does not restrict movement due to their finer count and elastic nature. The elastane fibres are very light, strong, and soft, and they offer better quality and comfort than rubber. Hence, elastane fibres are also used in undergarment fabrics like waistbands, support hose, brassieres, and briefs. In commercial clothing, elastane fibres are used in various apparel like hosiery, swimsuits, exercise wear, bodysuits, ski pants, disco jeans, skinny jeans, golf jackets, disposable diapers, gloves, slacks, leggings, socks, and many more. Elastane material is not used separately, it is used in a lower percentage in all the day-to-day garments. The elastane fibres are also used in compression therapy. They were used in garments to impart compression to the wearer either to give a comfortable feeling or to enhance the performance [17]. The common applications of elastane fibres are provided in Figure 1.5.

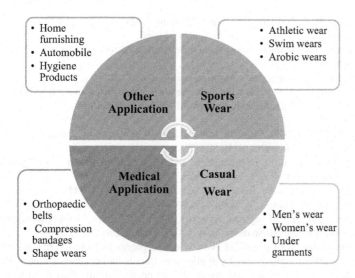

FIGURE 1.5   Applications of elastane fibre.

1.6   SUMMARY

This chapter provides a prelude to the readers regarding the elastane fibre production methods. The manufacturing of elastane is broadly divided into two segments in this chapter namely polymer interaction and spinning. The details of the dry spinning process with step-by-step production are outlined in the chapter. The physical and chemical characteristics discussed in the latter part were aimed to provide a clear understanding of elastane fibre properties.

REFERENCES

1. Spandex. (2021). How products are made – Forum, http://www.madehow.com/ Volume-4/Spandex.html (Accessed on July 2021).
2. Marc Reisch. (1999) "What's that Stuff? - Spandex". **77** (7). *Chemical & Engineering News*. (Retrieved 2018-12-06).
3. Anon. (1998). Joseph c. Shivers to receive the olney medal. American Association of Textile Chemists and Colorists, https://web.archive.org/web/20131203014041/http:// www.aatcc.org/awards/OlneyShivers.pdf (Accessed on July 2021).
4. Kadolph, S.J. (2006). *Textiles* (10th ed.). Pearson Prentice Hall. NJ, USA.
5. Sara, J.K., & Sara, M. (2016). *Textiles* (12th ed.). Pearson Publications, USA.
6. Fei, B. (2018). High performance fibers for textiles. In Book. In Menghe Miao & John H Xin (Eds.), *Engineering of high-performance textiles*. Woodhead Publishing Pvt Ltd, UK.
7. Jinlian Hu, Jing Lu, &Yong Zhu. (2008). New developments in elastic fibers. *Polymer Reviews*, 48(2), 275–301.
8. Anon. (2021). Fabric series: All about elastane, https://www.kleiderly.com/our-blog/ fabric-series-all-about-elastane (Accessed on July 2021).
9. Lewin, M., & Preston, J., (ed.). (1985). *High technology fibers*. New York: Marcel Dekker.

10. Bardhan, M.K., Sule, A. D., (2004). Anatomy of sportswear and leisurewear- Scope of Spandex fiber. *Man Made Text India*, 8(1), 81–86.

11. Elasthane, E.L. https://www.swicofil.com/commerce/products/elasthane/776/introduction (Accessed on July 2021).

12. Spandex. (2011). Spandex - Fashion, costume, and culture: Clothing, headwear, body decorations, and footwear through the ages. http://www.fashionencyclopedia.com/fashion_costume_culture/Modern-World-1980-2003/Spandex.html (Accessed on July 2021).

13. Haji, M.M.A. (2013). Physical and mechanical properties of cotton/spandex fabrics. *Pakistan Textile Journal*, 62(1), 52–55.

14. Arzu Marmarali. (2003). Dimensional and physical properties of cotton/spandex single jersey fabrics. *Textile Research Journal*, 73(1), 11–14.

15. Spandex Market Size. Share & trends analysis report by technology (wet-spinning, solution dry-spinning), by application (clothing, medical), by region (APAC, North America, MEA), and segment forecasts, 2020–2027, https://www.grandviewresearch.com/industry-analysis/spandex-market (Accessed on July 2021).

16. Spandex Fiber Market. (2021). By production processes (solution dry spinning, solution wet drying, and others), by applications (textile and healthcare) and regions (Asia Pacific, North America, Latin America, Europe, and Middle East & Africa) – Global Industry Analysis, Growth, Share, Size, Trends, and Forecast 2021–2028, https://dataintelo.com/report/spandex-fiber-market/ (Accessed on July 2021).

17. Mahapatra N.N., (2019). Clothing from Spandex fibres, Textile Value Chain, https://textilevaluechain.in/news-insights/clothing-from-spandex-fibres/ (Accessed on June 2021).

18. Gordon Cook, J. (2009). *Handbook of textile fibers, Part II – Manmade fibers*. New Delhi, India: Woodhead Publishing India Pvt Ltd.

19. Hughes, A.J., McIntyre, J.E., Clayton, G., Wright, P., Poynton, D.J., Atkinson, J., Morgan, P.E., Rose, L., Stevenson, P.A., Mohajer, A.A., & Ferguson, W.J. (1976). The production of man-made fibres. *Textile Progress*, 8(1), 1–156. http://dx.doi.org/10.1080/00405167608688984

20. Suresh Babu. (2019). *Textile adviser: elastane or spandex fibre.* Dorlastan: Lycra, https://www.textileadvisor.com/2019/10/elastane-or-spandex-fibre-lycra.html (Accessed on August 2021).

21. Hicks Jun, P.E.M., Craig, R.A., Wittbecker, E.L., Lavin, J.G., Ednie, N.A., Howe, D.E., Williams, E.D., Seaman, R.E., Pierce, N.C., Ultee, A.J., & Couper, M. (1971). The production of synthetic-polymer fibres. *Textile Progress*, 3(1), 1–108, http://dx.doi.org/10.1080/00405167108688988

22. Robert, J.D., & Caserio, M.C. (1977). *Basic principles of organic chemistry* (2nd ed.). Menlo Park, CA: W. A. Benjamin, Inc.

23. Roy, A.K. (2011). Dyeing of synthetic fibers. In Book. In M. Clark (Ed.). *Handbook of textile and industrial dyeing* (Vol. 2). Woodhead Publishing Ltd, UK.

24. Boliek, J.E., & Denney, S.A. (2002). Fibers, elastomeric. In Encyclopedia of Polymer Science and Technology (Vol. 6), 267–283. https://doi.org/10.1002/0471440264.pst129

2 Elastane yarn and fabric production

2.1 INTRODUCTION

Stretch fabrics are the type of fabric that can stretch or expand either in lengthwise or widthwise direction. The stretch fabrics can be classified as two-way stretch and four-way stretch fabrics. Two-way stretch fabrics are woven fabrics that can be stretched in any one direction, in contrast, four-way fabrics can be stretched both lengthwise and widthwise. But the knitted fabric can be stretched in all directions due to its structural arrangement. The stretch fabrics are mainly used for fitted apparel like sports clothing, fashion wears, and some medical wear. Based on the application, the stretch fabrics can also be termed as

i. Power stretch fabric
ii. Comfort stretch fabric

Whereas the power stretch fabrics mainly represent the fabric (both knitted and woven) that has more than 50% of extensibility in the form of garments. These apparels are mainly used by professional sports personals like swimmers, athletes, etc., who required a higher stretch and immediate recovery in the course of action. The second type, comfort stretch fabrics are the fabrics having stretch around 15–20%. These fabrics are commonly used in places where we require a very limited amount of extension. Examples for the comfort stretch applications are causal (everyday) apparel like stretch denim, sheer tights, etc. As the requirement for stretch apparels increases, the need for stretch fabrics also increased in the modern world. Hence, the elastane yarn consumption increased tremendously in the past decade. But, the produced elastane yarn in the form of filament cannot be used as such in many applications. For instance, bare elastane yarn can be used in knitted fabric production through plating. But for the development of woven fabric, elastane yarn is always used as a core yarn with cotton or other fibres as a sheath material. Hence, it is important and necessary to prepare the elastane yarn for the fabric production process. The previous chapter of this book detailed the production methods of the elastane fibre in both the chemical and physical aspects. This chapter aims to detail the different methods of elastane yarn production used in the industry. Elastane fibres are generally used in the following ways in their products as shown in Figure 2.1.

DOI: 10.1201/9780429094804-2

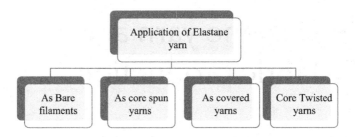

FIGURE 2.1   Application methods of elastane yarn in the textile.

2.2   ELASTANE AS BARE FILAMENT

As synthetic elastane yarns are stronger than natural rubber, the industry started using the bare yarn as such in the production of elastane based textile material. Unlike natural rubber yarn, elastane yarn does not require any cover to protect it due to its higher strength, abrasion resistance, and dyeing ability. The elastane fibres are commonly used as a bare yarn in foundation garments without any protection. Garments like swimwear, knitted tight-fit garments like tricot, lace, and all circular knits use bare elastane yarn. In the circular knitting process, the elastane yarns are directly used without any cover along with a ground yarn. The process of using the ground yarns and elastane filament together is known as plating. In knitted fabric, the bare elastane yarn is introduced into the structure in two ways namely (1) half plating and (2) full plating.

i. Full plating: In this method, bare elastane yarn will be used in all the courses of the ground yarn as shown in Figure 2.2a.
ii. Half plating: In this process, bare elastane yarn will be used in every alternative course of the ground yarn as shown in Figure 2.2b.

The use of bare yarn is always accompanied by a ground yarn, known as rigid yarn [1]. Both the elastane and ground yarn form a loop parallel in such a way that

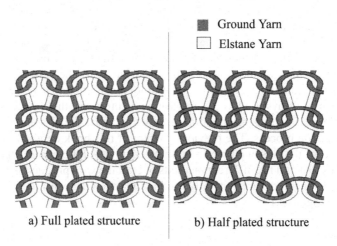

a) Full plated structure   b) Half plated structure

FIGURE 2.2   (a) Full-plated knit; (b) half-plated structure.

only the ground yarn is visible on the technical face of the fabric. The elastane yarn can be viewed on the reverse side of the fabric. It is important to have complete control over the elastane yarn feed rate, count, and feeding tension to obtain the desired results in the output fabric. The use of bare yarn enhanced the possibilities of producing thin fabric focusing on stretch and tight fit applications. Further, when compared to natural rubber, dyeing elastane with different colours is also one of the main reasons for the bare usage of elastane yarn. Similarly, elastane yarns are produced in finer counts than their predecessors and so it has wide applications.

2.3 ELASTANE AS CORE-SPUN YARN

Core-spun yarns are the type of yarn in which there are two different fibrous materials used to produce the yarn. In most cases, synthetic filaments (mono or multifilaments) will be used as a core element and the sheath components will be of staple fibres [2, 3]. The rationale behind the development of such composite or yarns with different components is to take advantage of the two different fibre materials in a single yarn [4]. The use of filaments in the core section of the yarn increases the strength of the yarn, whereas the cover fibres will give the properties of their nature (appearance, surface characteristics, and feel) to the wearer [5, 6]. Several researchers confirmed the superior properties of core-spun yarns than the 100% staple or filament yarns [7, 8]. Figure 2.3 represents the structure of elastane core-spun yarn.

As reported in Figure 2.3, to achieve the core-spun yarn structure, the normal spinning method cannot be used as such. The machinery needs to be modified based on the need as the existing or normal machines are designed to perform the drafting, twisting, and winding operations only for staple fibres. Hence, to produce the core-spun yarn with elastane core and staple fibre on the outside of the yarn, there should be some modifications performed in the spinning system. When the ring frame is considered, the following changes are necessary to develop elastane core-spun yarns [11].

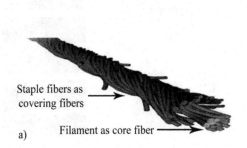
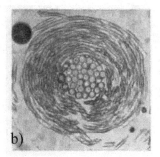

Staple fibers as covering fibers →

a)   Filament as core fiber →

b)

FIGURE 2.3 Elastane core-spun yarn cross-section. (a) Corn yarn length-wise structure [9] (reprinted under Creative Commons licence); (b) cross-section of nylon core and viscose sheath fibres (Images reprinted with permission from [10]).

1. Additional creel support to place the elastane yarn along with proper guides and tensioning mechanism.
2. The elastane filament from the creel should be positioned to feed in the nip of the front drafting rollers of the spinning system.
3. Hence, both the drafted staple fibres and filament passed into the front drafting roller.
4. The twisting mechanism twists both staple fibre and filament together and forms the core-spun yarn.

Though researchers attempted in changing the structure of the existing ring-spinning frame, the different characteristics of continuous elastane filament and staple fibres created various issues during the production of core-spun yarn. One such major issue is the slippage of staple fibres during the drafting process, relative to the filament yarn. This was noted as the major defect in the elastane core-spun yarn production. This defect was called as "strip back" or "barber pole" [12]. This defect will create a yarn with an improper covering of core yarn by sheath yarn. Imparting a higher amount of twists on the yarn is also known as one of the effective ways to control the defect despite its higher production cost. The normal method of incorporation of elastane yarn inside the core-spun yarn was reported by other researchers [13]. Hence, to overcome the problem, Sawhney et al. [14] developed a modification in the spinning frame as provided in Figure 2.4.

The elastane core-spun yarns can also be produced using the rotor spinning method with minor modifications as reported by Pouresfandiari et al. They have incorporated a filament feed tube with small ball bearings to avoid false twist. The arrangement presents the filament from package to rotor spinning machine through a suitable guide mechanism and tensioning device. The yarn is directly drawn through the rotor, doffing tube, and to take up roller as provided in Figure 2.5 [16]. The researcher reported that the system can handle variable overfeeding during spun yarn

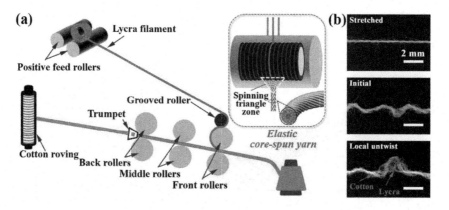

FIGURE 2.4   (a) Schematic diagram of elastic core-spun yarn production on a modified ring-spun frame; (b) appearances of an elastic core yarn sample in stretched and initial states, respectively (Reprinted under Creative Commons licence from [15]).

production [16]. It is important to consider a few other parameters for the better production of elastane core-spun yarn through rotor spinning [17], as given below,.

  i. The feed channel for filament must be on the axis of the rotor.
  ii. The diameter of the feed channel bore must be slightly larger than the filament diameter, which prevents unwanted air intake.
  iii. The doffing tube should be coaxially fitted on the other side of the rotor, with a delivery roller pair to wound the produced yarn.

During the production, the filament from the package is unravelled and sucked into the rotor section under pressure. During the spinning process, the staple fibres will be wrapped around the filament and passed through the doffing tube [17].

Nield and Ali reported another method to develop core-spun yarn from the rotor spinning process. As per this method, a feed tube must be positioned in line with the axis of the doffing tube and the elastane filament must feed in a controlled tension to avoid its misalignment in manufacturing. Improper tension in the filament will make the filament move away from its axis. During yarn production, the doffing tube is made to rotate in the opposite direction to develop a false twist to the yarn that is produced. When the filament passes through the rotor, the staple fibres wrap around the filament and drafter through the doffing tube. This was performed by the rotary movement of the rotor in the spinning machine. The false twist developed by the doffing tube helps in producing peel-off points. However, the yarns produced by this method had a serious limitation with counts [9]. The schematic diagram of core-spun rotor yarn production in provided in Figure 2.6.

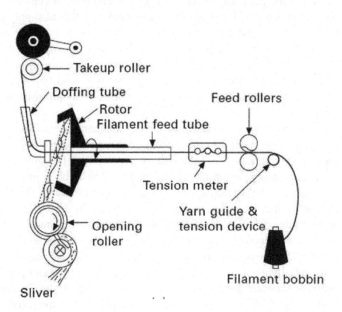

FIGURE 2.5 Elastane core-spun yarn production through rotor spinning machine (Reprinted with permission from [11]).

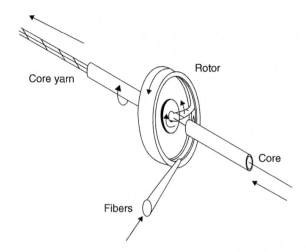

FIGURE 2.6   Rotor spinning for core-spun yarn production as reported by Neild and Ali [10] (Images reprinted with permission).

Another method for developing core-spun yarn from rotor spinning without any slippage between the core filament and sheath yarn was reported by Peter Artzt et al. In this method, a loop is formed in the core thread to delay the start of drawing rollers for a predetermined interval. The basic concept in this method is that the core filament is supplied at one speed and the core yarn formed at the rotor is drawn off at a slower speed. This speed difference of the core filament is shortened by twisting the fibres on it. In this method, the fibre feed tube and core thread supply tube were fixed in the opposite direction of the rotor plane. A high-density core yarn can be developed with lower slippage between the core and the wrapping fibres. The issue of relative slippage is eliminated in this method as the core thread is completely wrapped by the spun fibres. The apparatus also allows us to modify the rate of delivery of the core thread concerning rotor speed to produce core-spun yarn with different characteristics [18].

DREF spinning method is also used for the development of core-spun yarn. The completely opened fibres passed to the perforated drums and yarn is formed in the conventional DREF spinning. The strength is imparted by the binding of fibres together in the converging region of the two drums. The twist is imparted by the rotational movement of the frictional drum. Figure 2.7 represents the DREF spinning system designed for the development of core-spun yarn. In the DREF II spinning method, the deposition and the twist insertion on the yarn are generally obtained in the yarn tail. This is replaced by the core filament to produce a core-covered yarn. In the case of DREF III system, core filament (elastane) material is false twisted by the friction drums initially as the first step. In the second step, staple and parallel fibres from the ribbon are opened from sliver and allowed to wrap on the core filament as a result of cohesive force developed by the friction drum and strength of resulting yarn. The overall strength of the DREF yarn depends on the strength of the core filament and the number of sheath fibres that actively provide pressure on the core yarn

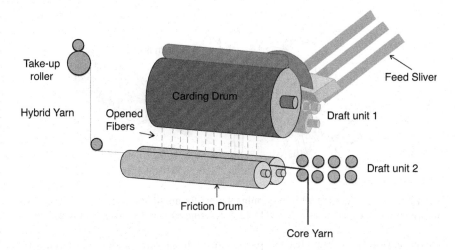

FIGURE 2.7 Core-spun yarn production from DREF spinning system (Reprinted with permission from [20]).

to form yarn structure. The physical and mechanical properties of the sheath fibre determine the effectiveness of the wrapping process [19].

There are few elemental changes that need to be incorporated to produce a core-spun yarn in a conventional DREF system. In Figure 2.7, it can be noted that apart from the regular draft unit, there was an additional draft unit (2) which was provided to deliver the filament core yarn from the package. On the other side, the feed sliver is drafted by drafting unit (1) which is common in conventional DREF spinning systems. Through carding, the opened individual parallel fibre allowed to fall in the friction drum, and these fibres are wrapped around the core filament as a sheath fibre. The cohesion and binding – in force developed by the friction roller, that ensures the formation of yarn. Developed core-spun yarn in the convergent region of drum is directly taken to the take-up roller as mentioned in Figure 2.7 and wound on a package [21]. Other researchers developed a method to produce multi-core-spun yarn from the DREF system with modification as shown in Figure 2.8 [22]. In this process, along with the normal drafting system (1) and friction rollers (3), the researchers added a second drafting system (6) as shown in Figure 2.8. In this method, there are two sets of core yarns used, yarn (5) represented as core yarn and yarn (4) mostly staple yarn represented as wrapper yarn. The two sets of yarns initially passed through the drafting system (6) via a specially developed trumpet (8) by the inventors. This helps in aligning the core (5) and wrapper (4) yarn correctly. Further, the inventors also added a pair of delivery or alignment rollers (9) in the system before the core components were delivered to the friction drums (3). At this point of juncture, the sheath fibres from sliver (2) are passed through the drafting system (1) and a carding roller to open the fibres (1a). The individual fibres fall over the core (5) and wrapper (4) yarn to form a sheath. As reported in the normal DREF system, due to the friction and rotary movement of the friction rollers (3), the core-spun yarn is formed (7) at the convergent region of the drum. The developed yarns (7) are continuously withdrawn

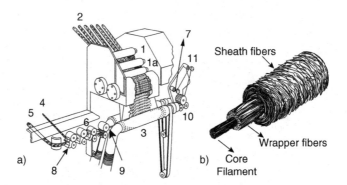

FIGURE 2.8 (a) Modified friction spinning system for multiple core-spun yarn production; (b) yarn structure expected as reported by [22] (Reprinted under Creative Commons licence).

by a delivery roller set up (10) and passed through set yarn guides (11) and allowed to wound on a package.

2.4 ELASTANE AS COVERED YARN/WRAP YARN

Wrap yarns are composite yarns that typically consist of core fibres, either in the form of twisted or twistless and those were bound by a yarn or continuous filament. The wrapped yarn is another commonly used method for the production of elastane-covered yarns. Though wrap spinning methods are not new, they gained a lot of attention for elastane yarn production in recent times. The main demerit of wrap yarn is its bulkier structure and the advantages are higher strength along with a good handle. Similarly, the wrapping yarns also can be of filament type or staple yarn type [23]. Figure 2.9 represents the schematic structure of wrap yarns.

The most common method that is used to produce the wrap yarn are discussed in this section for the production of elastane yarns.

2.4.1 HOLLOW SPINDLE TECHNOLOGY

Out of several existing wrap yarn production methods, the hollow spindle method is one of the commonly used methods. The outline of the hollow spinning apparatus is provided in Figure 2.10. In a hollow spindle system, the roving from the creel is passed to a drafting system with four or five rollers. Once the roving passed the drafting zone (1), it enters into the hollow spindle zone (2). Where the drafted yarn passes down into the hollow spindle, where it is false twisted. The hollow spindle also contains a spindle with continuous filament to wrap it around the core fibre. Based on the machine type, the false twist unit may be fitted either on the top or bottom side of the hollow spindle unit (2). The type of false twisting element that is used in the hollow spinning system is the pin type. When the drafted yarn enters into hollow spindle the filament gets wrapped but at the same time as the filament pirn also rotates with the hollow spindle, there is no twist imparted on the filament. After the completion of the wrapping process, the produced yarn is allowed to winding zone (4) via a pair of delivery rollers (3) [25–27].

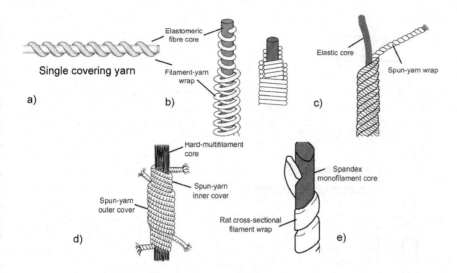

FIGURE 2.9   Different types of wrap methods for elastane core yarns. (a) Single covered (filament) wrap yarn, (b) double covered (filament) wrap yarn, (c) spun yarn wrap with single yarn, (d) double cover with spun yarn, and (e) elastane covered with filament (Reprinted with permission from [24]).

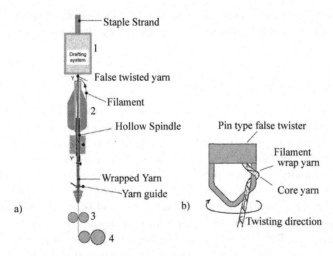

FIGURE 2.10   (a) Schematic of hollow spindle technology to produce wrap yarn and (b) false twist device used in hollow spindle machine (Reprinted with permission from [27]).

2.4.2   SELFIL SPINNING METHOD

Selfil method of wrap yarn production is used for both staple core and filament core. In this method, two filaments are wrapped on the core fibre using a modified Repco self-twister. The self-twist units fixed with this machine helps in wrapping the wrapper fibres around the core yarn. In the cases of two wrapper filaments, both the filaments are wrapped in the opposite direction. The resultant yarns are double-covered

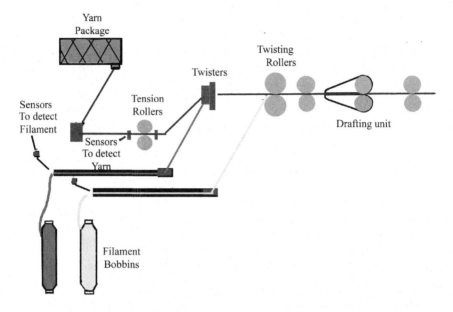

FIGURE 2.11 SELFIL wrap yarn production method [23].

yarn and are also sometimes termed as Selfil yarns. The schematic of the Selfil spinning process is provided in Figure 2.11 as reported by Brydon and van der Merwe (1986) [23].

2.4.3 SELF-TWIST SPINNING METHOD

Self-twist spinning method of wrap yarn production consists of Repco wrapped corespun yarn system (RWCS), Repco wrapped spun yarn (RWS), and Repco doublewrapped core-spun yarn system (RDWCS) [1]. In this, RWCS is commonly used for the staple fibre or filament core yarn wrapped with a filament yarn. Though this system was initially produced for the development of mohair yarns, later it was adapted for the common application. Out of these systems, the RWS system uses only the staple core yarns and RDWCS uses both staple and filament core. Additionally RDWCS system uses two filament wrapper yarns to develop the structure. The schematic illustration of RWCS is provided in Figure 2.12 as reported by Mahmoudi [28]. Other researchers detailed the other types of wrap yarn production methods and different yarn structures in their research [23].

2.4.4 MURATA VORTEX SPINNING

In this method, air-jet nozzles are used to create the wrapped yarn. This method is capable of producing ring-spun yarn like yarns with higher tenacity due to more number of wrapper fibres. The schematic structure consists of an air jet device that consists of a nozzle block and an injector, followed by a hollow spindle and a guide. The drafted sliver is passed to the air-jet nozzle and got twisted by swirling air

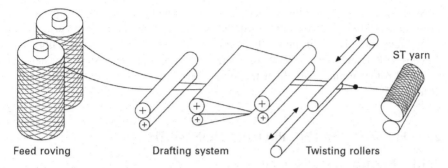

FIGURE 2.12   Schematic illustration of self-twist spinning system (Reprinted with permission from [28]).

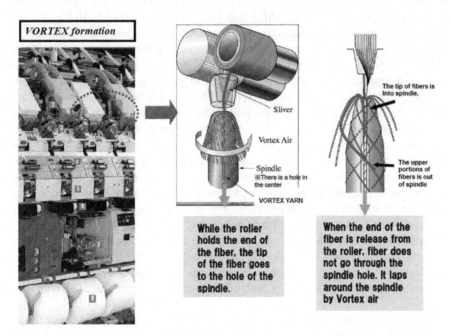

FIGURE 2.13   Mechanism of vortex yarn formation (Reprinted under Creative Commons licence from [29]).

towards the upward direction. Once the yarn enters into the hollow system, some fibres are kept open while it is delivered from the front drafting roller, due to the air jet movement helps in complete yarn formation by wrapping the core fibre [29]. The mechanism of yarn formation through vortex spinning is provided in Figure 2.13. For the production of elastane core-spun yarn, researchers reported that the system must have an elastane feed device along with a normal set up which was detailed in their study [30]. In this method, both elastane filament and staple fibres were merged at the nip point of the front rollers of the drafting unit. Though the elastane yarn is stretched under the drafting zone, the machine manufacturer claims that it is not twisted during manufacturing. Hence, elastane filament not only escapes from damages but also aligns to the centre of the yarn produced due to the basic principle of air-jet machines. This eliminates the slip of core yarn from the covering fibres and reduces subsequent fabric defects. Further, the vortex spinning system is faster than ring and rotor spinning in which the staple yarns can be spun at 400 metres per minute [30].

2.5 ELASTANE FABRIC PRODUCTION METHODS

2.5.1 KNITTED FABRIC PRODUCTION

In knitted fabric production, elastane yarns are used as such without any cover around its structure. Elastane yarns are usually processed with some other ground yarn due to their higher extensibility and lower breaking strength. Hence, the use of ground yarn is always necessary to overcome its limitations in the production and application phase of elastane fabric. As detailed in the previous section of this chapter, blending elastane yarn with natural or synthetic textile yarn is known as plating. It represents the formation of a single loop with two (one ground yarn and another elastane yarn) yarns, in which one thread will be reflected on the face side of the fabric and the other on the backside. The elastane yarn count, feed ratio, and tension play a vital role in the fabric properties. Hence, it is important to control these parameters efficiently. To control these properties, elastane yarns always pass through guide rollers, which reduces the friction and provides positive feeding to the knitting machine [31]. The positive feeding mechanisms represent the unwinding of elastane bobbin through a controlled motion and the yarn passes through a set of electrical sensors or stopping devices. Through a plating roller, the elastane yarn is directly fed to the needle for fabric formation. As elastane feeding is one of the main factors, several manufacturers had launched their products into the market to provide positive control to feeding. Figure 2.14 represents the elastane feeding roller developed by BTSR (Best Technologies Study and Research, Italy) [32]. Figure 2.15 represents yarn passage scheme of both ground yarn and elastane yarn in the knitting machine.

The ground yarn is also delivered from a positive storage feeder to the needle through the carrier plate. The yarn also passes through the guide holes in the carrier plate and passes to the needle. The number of yarns that pass through the guide hole is optional as per the fabric requirements. Elastane yarn also passes through the sensors and carrier plate. Both the yarns used to have a separate guide slot to present the ground yarn and elastane to the needle [33].

The addition of feed rollers to elastane yarns provide better control and smooth feeding during the halt and restarting of the knitting machine. Further, based on the requirements, these devices can be adjusted with different yarn tensions. The devices were able to provide a wide range of tensions from 0.5 g/Tex to 50 g/Tex. The sensor in the devices (black colour box in Figure 2.14) eliminates the yarn twisting and ensures the delivery of the flat yarn. Some latest feeding devices are also available with yarn breakage, yarn feed rate, and the total yarn length delivered measuring sensors.

FIGURE 2.14 Bare elastane feed roller for knitting machine, developed by BTSR, Italy.

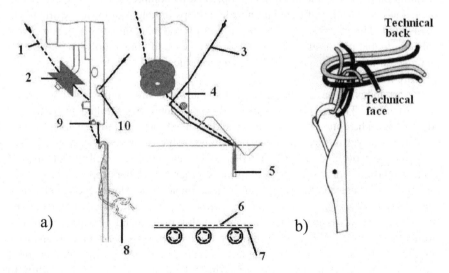

FIGURE 2.15 Yarn passage outline for (a) elastane yarn (dotted line); (b) ground yarn; (c) yarn arrangement in fabric (black colour – ground yarn) (1, spandex yarn; 2, change of direction roll; 3, cotton yarn; 4, carrier plate; 5, needle; 6, spandex yarn plated on the backside of the fabric; 7, cotton yarn at the front side of the fabric; 8, loops formed by needles; 9, spandex yarn feed slot; 10, guide hole for cotton) [34] (Reprinted under Creative Commons licence).

2.5.2 Woven fabric production

Bare elastane yarn is not used in the weaving industry. In woven fabric, elastane yarn is used in the form of core-spun yarn. The core-spun elastane yarn is either used as weft yarn or in warp yarn or on both sides based on the requirement of endues. This helps in achieving the properties of outer sheath yarn along with the stretch properties. Similar to the knitted fabric, in the case of woven also the elastane yarn linear density, feed ratio, and type of elastane yarn are used to decide the fabric properties including the weight and stretchability.

(i) Elastane yarn in weft direction

The most common method of using elastane yarn in the weft direction is through a rapier or air-jet weaving machine. As the rapier weaving machine mechanically carries the weft yarn, it is commercially used to produce most of the elastane fabric. The weft stretch fabrics like stretch denim and stretchwear were produced by rapier machines commercially. Comparatively in the case of air-jet looms, a higher amount of problems was noticed due to the different weft carrier mechanisms. The weaving ability of elastane core-spun yarn in the air jet loom was measured by Simon De Meulemeester et al. [35]. They have measured the wearability of elastane core-spun, core twist, and air-covered yarn in air-jet looms. During the weaving of air-covered yarn, several issues like delay in weft insertion (irrespective of the air pressure applied), higher amount of sheath yarn clogged on the nozzle, and double insertion of yarn causing double pick and thickness on the fabric as a recurring defect were reported. To avoid such issues, several machine manufacturers developed a mechanical clamp on the moveable main nozzle operated by compressed air. The clamp can be fixed at the entrance of the main nozzle, it helps in holding the weft while being not inserted. This also helps in the machine stop and start-up situations and eliminates the double pick issue [36]. Whereas in the case of core-spun yarn, the sheath got removed frequently compared to other yarn due to the air jet pressure as shown in Figure 2.16.

The air index tester results showed that the yarns are having good weavability (below 3%). The results showed that lower count yarns resulted in higher air index values. The core-spun and air-covered yarn exhibited similar air index values and core twist yarns showed a lower air index value. This represents that lighter yarns can be woven faster [37]. Further to evaluate the wearability potential of filament (elastane) or cotton core-spun yarn, other researchers reported a simulation method. Webster is the instrument that can able to develop a cyclic extension, axial abrasion, flexing, and bending, excluding beat-up and yarn entanglements by simulating the real-time weaving process. In this instrument, 15 yarns were held parallel and the tension applied. The results were noted up to the break of ten yarns and after the break of every yarn, the load is reduced to 1/15th of the total or initial load. This is to maintain uniform tension throughout the experiment. Especially to measure the core-spun yarn weavability, once the core yarn is visible it was considered as a break. Through this instrument, researchers analysed the weaving ability of core-spun cotton and polyester filament yarn and reported cotton core yarn as superior one [37].

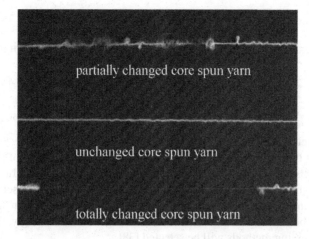

partially changed core spun yarn

unchanged core spun yarn

totally changed core spun yarn

FIGURE 2.16   Core sheath removal of core-spun yarn during air-jet insertion as weft in air-jet loom (Reprinted under Creative Commons licence from [35]).

(ii)   Elastane yarn in the warp direction

Use of elastane yarn in the warp direction is generally required to develop a four-way stretch fabric. However, introducing elastane thread in the warp direction is typically a complex process due to the warp beam preparation. As similar to the normal warp yarn preparation, all the processes need to be performed from warping, drawing to denting in. It is also paramount to maintain uniform tension on the warp thread during this process as it has a serious impact on the final fabric quality. Hence, due to the labourious nature, and higher cost, most of the time weft direction is preferred as a suitable method for stretch woven fabric. Other than usual processing machines, no additional equipment is required to perform this activity.

2.6   PROCESSING OF ELASTANE FABRIC

Developed elastane fabric both from the weaving and knitting process will undergo a set of processing steps like conventional fabric. The process sequence of the elastane fabric is provided in Figure 2.17.

2.6.1   RELAXATION

It is necessary to relax the elastane fabric always to lower their energy state. Before taking it to the next process, both the woven and knitted fabrics need to be relaxed to reduce the residual stress caused by the fabric manufacturing process. The relaxation process increases the dimensional stability of the elastane fabric. It is also important to perform a heat setting process to prevent the creping of yarns, structural distortion, and uneven dyeing and finishing. In the case of woven elastane fabric, the relaxation process increases the bulkiness of the fabric. This process ultimately reduces the fabric width in the case of woven fabric. Hence, it is important to optimize the proper relaxation so that the resultant bulkiness will yield desired end use requirements like

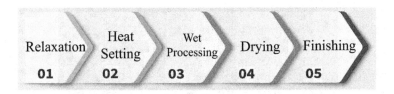

FIGURE 2.17   Wet processing stages involved in the elastane fabric processing.

dyeability, stretch, and recovery. For both the woven and knitted fabric, the relaxation process helps in finding the maximum weight and minimum width of the material. To measure this phenomenon, the boil-off method is commonly used. In this process, a full-width sample to a unit length is boiled along with a detergent or scouring agent. The boiled samples are then dried and measured for changes in dimension. These changes will be considered as the maximum change and accordingly, the temperature and processing methods will be selected [38].

In the case of woven fabric, a higher temperature will result in more width loss. Hence, controlling the processing temperature is very important in the case of dimensional shrinkage. If the fabric was not relaxed completely, then the fabric will form creases during the subsequent creasing and result in a permanent fabric defect. Hence, it is necessary to complete the relaxation process correctly. The fabric can be relaxed in different ways like steaming in a soft flow jet or pre-scouring or through padding mangle [39].

2.6.2   HEAT SETTING

Heat setting of elastane fabric can be done either before or after the scouring, bleaching, and dyeing. But sometimes, the heat setting process may result in a colour change or yellowing. If the heat setting is performed on greige goods, the oil, wax, and spinning oils may get permanently deposited on the fabric. This cannot be removed in the subsequent scouring and bleaching process. Hence, to avoid such issues, the fabrics must be treated with hot water and a wetting agent before the heat setting process. The heat setting process will be performed in the range of 182–196⁰C. During the heat setting process, the fabric width should be maintained 5%–15% wider than the required width as there will be a small stretch on the fabric even after heat setting [40]. In the case of open width fabric process, before heat setting, there should be an even distribution of moisture in the fabric. If it is stored for a long time, then it should be rewetted before the heat setting process to avoid uneven heat application. This will reflect in patchy dyeing and finishing in the subsequent process [39]. The tension applied on the fabric on a pin stenter frame must be uniform and optimum, a higher stretch will result in edge curing and high shrinkage. In the case of woven fabric with weft stretch, normal stenter is enough as similar to the knitted fabric. However, when the warp stretch fabrics are used, the tension must be applied in the warp direction using a jig or continuous range [39].

For knitted fabric with a tubular structure, the heat setting is performed in an autoclave with a help of high-pressure steam. In this process, it is important to autoclave an entire batch to avoid shade variations in the dyeing process. After the heat setting process, the fabric must be cooled and passed to the next process to avoid shade differences due to temperature differences. The heat setting process not only affects the wet processing quality but also has a strong influence on the mechanical properties of the fabric produced. Technically when the heat is applied to the elastane yarn, the inter molecules are rearranging themselves due to the molecular bond breakages. Most of the time, the elastane fabric is stretched and passed to the heat setting process, this process reduces the retractive force of the fabric due to molecular structure rearrangement. Further, the fibre fineness significantly reduces with the heat setting of core-spun elastane yarn [41]. The mechanical properties of heat-set fabric were analysed for different process routes, namely, heat set and pre-treatment, pre-treatment and heat setting, and also a fabric without heat setting. The results showed that the heat setting process significantly impacts mechanical properties like elongation, growth, tensile strength, and tear strength. But in the case of total hand value, all the fabrics showed a similar value around 3.5–3.7 [42].

On analysing the properties of cotton core-spun elastane fabric properties, research showed a significant impact of heat setting on fabric stretch properties. The heat setting temperature highly influences the elastic portion of the elastane yarn and provides a compact nature and stability. This, in turn, reduces the growth, stretch, and recovery properties of the elastane woven fabrics [43]. Heat treatment before any processing was highly recommended by the researchers due to various quality-related issues. It also reported that higher temperature treatment may spoil the elasticity and characteristics of the elastane fabric [44]. When the fabric width extension is higher along with overfeeding, it decreases the fabric length way shrinkage and results in higher stability. But, at the same time, a lower feeding results in higher width way shrinkage on laundry [45]. In another study, the researcher showcased the effect of heat setting temperature on the dynamic work recovery properties of the elastane plated knitted fabric. The results showed that the heat setting temperature had influenced the dynamic work recovery properties in both the course and wales wise direction of the knitted fabric [34].

2.6.3 WET PROCESSING

There are not many differences in the dyeing of elastane fabric and normal fabric. In the case of both woven and knitted fabric, the fabric must be kept as flat as possible throughout the process to avoid uneven tension. When the fabric is passed through any machine, care must be given to the widthwise extension or filling stretch of the fabric. Further, the machine with lower pressure should be selected for dyeing like a soft flow dyeing machine. Use of bleaching agents such as hydrogen peroxide is recommended as it does not affect the elastane fabric compared to chlorine-based bleaches. The chlorine-based bleaching agents may create yellowing problems in the elastane fabrics [34]. The dyed fabrics must pass through a centrifugal extraction process for drying over the padding mangle vacuum applications [39]. If the heat setting is performed in between, then there may be drying is required before passing it to the heat setting.

For general cotton/elastane fabric dyeing, the process sequence remains similar to cotton. In some cases, as the elastane filament is inside the cotton sheath yarn, dyeing will be performed without affecting the elastane core, termed as elastane reserve dyeing. If the elastane yarn needs to be dyed, like in the case of bare filament usage, then 1:2 metal complex monoazo dyes will be used. In the case of simultaneous dyeing of both cotton and spandex, metalized or milling acid dyes are used followed by a direct dye at a higher temperature [46]. The common dyeing sequence of the elastane fabric (woven or knitted) can be mentioned as provided in Figure 2.18 [38].

2.6.4 DRYING

As already heat setting is completed, it is always recommended to go with minimum temperature for drying the elastane fabric. Higher temperature drying may result in the yellowing of the fabric. Woven fabrics may be dried in open-width stenter and tubular fabrics can be dried in relaxation dryers.

2.6.5 FINISHING

Both the mechanical and chemical finishing of elastane fabric can be performed. Care should be taken only in the case of finishes that apply high temperatures. Whereas in the case of chemical finishing, the fabric is compatible with all the finishes that are used for cotton fabric. While applying mechanical finish, it is important to maintain

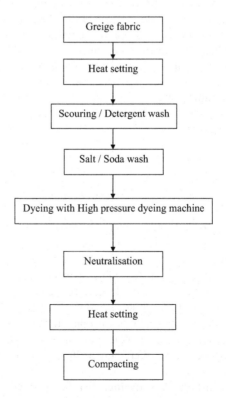

FIGURE 2.18   Stages of cotton/elastane fabric dyeing process.

the applied pressure at a minimal level. For woven fabrics, a sanforizing finish is necessary to control the length-wise shrinkage and also to remove the wrinkles from the edges. For both knitted and woven fabric, the compacting process is generally used to physically modify the yarn geometry in the fabric. In woven fabric, compacting process arranges the weft yarns closer and shrinks the fabric width. In the case of knitted fabric, the rearrangement of loop shape will result in the reduction of dimensional changes caused by different stretch forces applied during the wet processing stages [47, 38]. The detailed processing sequence of the elastane knit and woven fabrics can be found in the previous literature [38].

2.7 SUMMARY

Various modes of elastane yarn application in textile fabric manufacturing were detailed from the yarn manufacturing and fabric manufacturing stage. The yarn manufacturing process outlined the most common spinning methods adapted in the core-spun elastane yarn-production process. Fabric manufacturing details the knitted (bare elastane and core-spun yarn) and woven (elastane core-spun yarn) fabric production with elastane filament. The fabric-processing steps involved in the treatment of elastane fabric are also outlined with special emphasis on the heat setting process.

REFERENCES

1. Spencer, D. (2001). *Knitting technology* (3rd ed.). London: Pergamon Press.
2. Merati, A.A., Konda, F., Okamura, M., & Murui, E. (1998). Filament pre-tension in core yarn friction spinning. *Textile Research Journal*, 68(4), 254–264.
3. Balasubramanian, N., & Nerurkar, S.K. (1974). Strength and elangation of cotton, polyster and polyester/cotton blends at different stages of manufacture. *Textile Research Journal*, 10(44), 106–111.
4. Akankwasa, N.T., Qasim Siddiqui, Edwin Kamalha, & Llyod Ndlovu. (2013). Cotton-elastane ring core spun yarn: A review. *Research Review in Polymer*, 4(4), 127–137.
5. Jeddi, A.A., & Merati; A.A. (1997). A study on the structural and physical properties of the cotton-covered nylon filament core-spun yarns. *Journal of the Textile Institute*, 88(1), 12–20.
6. Suk, S.K., & Lee; J.K. (1978). Astudy on the physical properties of the core spun yarn. *Journal Of Korean Fibre Society*, 15(1), 23–31.
7. Lee, J.K. (1976). A study on the core-spun yarn. *Journal of Korean Fiber Society*, 13(2), 36–41.
8. Rameshkumar, C., Kavinmurugan, P., Manojkumar, B., & Anbumani; N. (2008). Influence of core component on the properties of friction spun yarns. *AUTEX Research Journal*, 8(4), 106–110.
9. Yılönü, S., & Ünal, B.Z. (2018). Investigating the effects of core spun yarns on the quick-dry property of towels. *Fibres & Textiles in Eastern Europe*, 3(129), 46–51. DOI: 10.5604/01.3001.0010.7772
10. Nield, R.; & Ali, A.R.A. (1977). 26—Open-end-spun core-spun yarns. *The Journal of The Textile Institute*, 68(7), 223–229. doi:10.1080/00405007708631386
11. Alagirusamy, R., Fangueiro, R., Ogale, V., & Padaki, N. (2006). Hybrid yarns and textile preforming for thermoplastic composites. *Textile Progress*, 38(4), 1–71. DOI: 10.1533/ tepr.2006.0004

12. Babay, A., Helali, H., & Msahli, S. (2014). Study of the mechanical behaviour of the elastic-core-spun yarns. *The Journal of The Textile Institute*, 105:7, 701–710, DOI: 10.1080/00405000.2013.844420

13. Oxenham, W., Lawrence, C.A., East, G.C., & Jou, G.T. (1996). The physical properties of composite yarns produced by an electrostatic filament-charging method. *Journal of Textile Institute*, 1(1): 78–96.

14. Sawhney, A.P.S., Ruppenicker, G.F., Kimmel, L.B., & Robert, K.Q. (1992). Comparison of filament-core spun yarns produced by new and conventional methods. *Textile Research Journal*, 62(2), 67–73.

15. Wang, Y., Qiao, Q., Ding, Z., & Sun, F. (2021). Strain-dependent wicking behavior of cotton/lycra elastic woven fabric for sportswear. *e-Polymers*, 21(1), 263–271. https://doi.org/10.1515/epoly-2021-0030

16. Pouresfandiari, F., Fushimi, S., Sakaguchi, A., Saito, H., Toriumi, K., Nishimatsu, T., Shimizu, Y., Shirai, H., Matsumoto, & Y.I., Gong, H. (2002) Spinning conditions and characteristics of open-end rotor spun hybrid yarns, *Textile Research Journal*, 72(1), 61–70.

17. Vaclar Rohlena et al., (1975). *Open-end spinning, textile science and technology*. Amsterdam Oxford, New York: Elsevier Scientific Publication Company, p. 99.

18. Peter Artzt, R., Albert Bausch, M., Gerhard Eigbers, E. (1976). All of Germany – method and apparatus for the manufacture of core yarn in an open end spinning device. United States Patent, May 3, 1976.

19. Tyagi, G.K. (2010). Yarn structure and properties from different spinning techniques. In Book. In Lawrence, C.A. (Ed.), *Advances in yarn spinning technology* (pp. 119–154). England: Woodhead Publishing Series in Textiles.

20. Vijay, G., Ramasamy, A., Apurba, D., & Dinesh, K. (2019). Influence of various forms of polypropylene matrix (fiber, powder and film states) on the flexural strength of carbon-polypropylene composites. *Composites Part B* 166, 56–64.

21. Klein, W. (1993). *New spinning system, short staple spinning series* (vol. 5). Manchester, England: The textile Institute.

22. Montgomery, T.G., & Martin, W.G. (1990). Corespun yarn friction spinning apparatus and method. *European Patent* 375(112), A2.

23. Brydon, A.G., & van der Merwe, J.P. (1986). Wrap spinning: principles and development (vol. 73), South African Wool and Textile Research Institute of the CSIR. SAWTRI Special Publication, https://vdocument.in/wrap-spinning-principles-and-developmentpdf.html

24. Alagirusamy, R., & Das, A. (2015). Conversion of fibre to yarn: an overview. In Book. In Rose Sinclair (Ed.), *Textiles and fashion materials, design and technology*. Cambridge: Woodhead Publishing. http://dx.doi.org/10.1016/B978-1-84569-931-4.00008-8

25. Javier Massó de Rafael. (2015). Wrap yarn technology, fundamentals and prototype design (Bachelor's thesis), Université de Haute Alsace, ENSISA. https://upcommons.upc.edu/bitstream/handle/2117/81913/WRAP%20YARN%20TECHNOLOGY%20Javier%20Mass%C3%B3.pdf

26. Carl, A.L. (2013). *Fundamentals of spun yarn technology*. Boca Raton, FL: CRC Press.

27. Peter, R.L. (2003). Staple systems and modified yarn structures. In Book. In Peter R. Lord (Ed.), *Handbook of Yarn Production* (pp. 260–275). England: Woodhead Publishing Series in Textiles. https://doi.org/10.1533/9781855738652.260

28. Mahmoudi, M.R. (2010). Self-twist spinning. In Book. In Lawrence, C. A. (Ed.), *Advances in Yarn Spinning Technology* (pp. 365–389). England: Woodhead Publishing Series in Textiles. doi:10.1533/9780857090218.2.365

29. Gizem Karakan Günaydin & Ali Serkan Soydan. (March 1st 2017). *Vortex spinning system and vortex yarn structure, vortex structures in fluid dynamic problems*. Hector Perez-de-Tejada: IntechOpen. https://www.intechopen.com/chapters/53651

30. Hüseyin Gazi Ortlek & Sukriye Ulku. (2007). Effects of spandex and yarn counts on the properties of elastic core-spun yarns produced on murata vortex spinner. *Textile Research Journal*, 77(6), 432.
31. Saber Ben Abdessalem, Youssef Ben Abdelkader, Sofiene Mokhtar, & Saber Elmarzougui. (2009). Influence of elastane consumption on plated plain knitted fabric characteristics. *Journal of Engineered Fibers and Fabrics, 4*(4), 30–35.
32. BTSR introduces Rolling Feeder for bare elastane feed. (2011). https://www.knittingindustry.com/btsr-introduces-rolling-feeder-for-bare-elastane-feed/ (Accessed on November 2022).
33. Laycock, G.H. (2006). INVISTA North America, method to make circular knit elastic fabric comprising spandex and hard yarns. US Patent No.7117695.
34. Senthilkumar, M. (2014). Dynamics of elastic knitted fabrics for tight fit sportswear. Anna University Ph.D., thesis. http://hdl.handle.net/10603/15503
35. Simon De Meulemeester, Lieva Van Langenhove, & Paul Kiekens. (2009). Study of the weavability of elastane based stretch yarns on air-jet looms. *AUTEX Research Journal*, 9(2), 54–60. http://www.autexrj.com/cms/zalaczone_pliki/0315_1.pdf
36. Picanol Movable clamp, www.piconal.be, picanol.be/sites/default/files/2020-12/WU_ClampMovableMainNozzle.pdf (Accessed on November 2022).
37. Behera, B.K., & Joshi, V.K. (2007). Weavability of core-spun dref yarns. *Indian Journal of Fibre & Textile Research*, 32, 40–46.
38. Mathews, K. (2019). *Pretreatment of textile substrates, handbook of textile processors series*. New Delhi: Woodhead Publishing India Pvt. Ltd.
39. Technical Bulletin. (2003). Wet processing of cotton/spandex fabric. Cotton Incorporated, North Carolina, USA. https://www.cottoninc.com/wp-content/uploads/2017/12/TRI-3012-Wet-Processing-of-Cotton-Spandex-Fabric.pdf
40. Technical Bulletin. (2003). Dyeing and finishing of cotton/spandex circular knits. Cotton Incorporated Technical Services Information, Raleigh, NC, USA. https://www.cottoninc.com/wp-content/uploads/2017/12/TRI-3012-Wet-Processing-of-Cotton-Spandex-Fabric.pdf.
41. McIntyre, J.E. (2005). *Synthetic fibres: nylon, polyester, acrylic, polyolefin*. Cambridge, England: Woodhead Publication.
42. Pannu, S., Ahirwar, M., Jamdagni, R., et al. (2020). Role of heat setting and finishing treatment on mechanical properties and hand behavior of stretch fabric. *Journal of Textile Engineering Fashion Technology, 6*(5), 169–177. DOI: 10.15406/jteft.2020.06.00247
43. Shariful Islam, Shaikh Md. Mominul Alam, & Shilpi Akter. (2018). Identifying a suitable heat setting temperature to optimize the elastic performances of cotton spandex woven fabric. *Research Journal of Textile and Apparel*. 22(3), 260–270. https://doi.org/10.1108/RJTA-01-2018-0002
44. Islam, S. (2019). Attaining optimum strength of cotton-spandex woven fabric by apposite heat-setting temperature. *Journal of Institutional Engineering India Service C,* 100, 601–606. https://doi.org/10.1007/s40032-018-0478-y
45. Ahsan Nazir, Tanveer Hussain, Aisha Rehman, & Affan Abid. (2015). Modelling heat-setting of cotton/elastane knitted fabrics for optimum dimensional stability. *Journal of Textile and Apparel Technology and Management, 9*(2), 1–12.
46. Anon. (2021). How can I dye elastane? http://www.pburch.net/dyeing/FAQ/spandex.shtml#:~:text=Spandex%20will%20be%20ruined%20by,Flow%20or%20Dharma%20Pigment%20Dye (Accessed on November 2022).
47. Hassan, M.B. (2005). *Effects of mechanical and physical properties on fabric hand.* Cambridge: Wood Head Publications.

3 Characteristics of elastane fabrics

3.1 INTRODUCTION

Knitted textiles are commonly used in tight-fitted garments due to their higher elastic nature than the woven fabric. The increasing trend towards the fitted silhouette attracts the customer to the knitted fabric. Due to their higher level of comfort at different applications like sportswear, leisure wear, casual, and innerwear, stretchable garments are always a part of the market. However, to increase the elasticity of the garment above a certain level, it is necessary to include elastane into the yarn or fabric structures, which is common for both knitted and woven fabrics. The inclusion of elastic yarns into the fabric structure increases the elasticity of any type of fabric. Hence, it is able to create a better fit for the body and has attracted a lot of market potential in recent years. Though elastane was introduced for tight wear like swimsuits, corsets, athletic wear, etc., due to the increase in awareness, it is quite common in fashionable and also in functional wear. Meanwhile, the elastane-added fabric also showed a few adverse characteristics. The addition of elastane in the fabric structure increases the fabric thickness and the weight of the fabric per square metre. Due to their higher tension under constant draw ratio, it elastane increases the fabric courses per centimetre, wales per centimetre, and thickness which leads to lower air permeability of the fabric. Further, it reduces the pilling grade and spirality of the fabric [1].

In this aspect, this chapter consolidates the effect of elastane inclusion on the dimensional, stretch and recovery, bagging, dynamic elastic recovery, physical, moisture, thermal, and other mechanical properties of the textile material.

3.2 DIMENSIONAL PROPERTIES

Dimensional properties of the fabric represent the structural stability at repeated use and further processing like washing and ironing. Compared to woven fabric, knitted fabrics often have lower dimensional stability due to their structure. The addition of elastane yarn in the fabric structure will significantly modify the dimensional characteristics of both woven and knit structures. The dimensional property of the fabric is greatly influenced by the fabric parameters like warp and weft yarn density (woven), loop density (knitted), cover or tightness factor, thickness, loop length, etc. A comparative analysis between 100% cotton fabric and cotton/elastane (95%/5%) fabric revealed the effect of elastane addition on various dimensional parameters. Results showed that the addition of 5% elastane content in the fabric significantly reduces the loop length of the fleece, single pique, interlock, 1×1 rib, and single jersey fabric. Out of all the samples analysed, a higher dispersion rate was noted with the fleece

DOI: 10.1201/9780429094804-3

fabric [2]. The course per centimetre and wales per centimetre of the selected knitted fabric were considered, the inclusion of 5% elastane in the knit structure significantly modified the yarn density in both direction. Out of the selected structures, the highest wales per centimetre was noted with interlock structure and the lowest noted with fleece structure. In the case of course density, a higher course per centimetre was found in single jersey, single pique, and fleece structure and reduced in interlock and 1×1 rib structure [2]. In the case of loop density, single jersey, single pique, and fleece fabric showed an increase in the loop density after the inclusion of 5% elastane in its structure, whereas in the case of interlock and 1×1rib structure it was noted as lower than other structure. Concerning the tightness factor, a higher tightness was noted with the interlock structure. The addition of 5% elastane in structure increased the tightness in single pique and two-thread fleece, whereas in the case of interlock, single jersey, and 1×1rib knitted fabrics the tightness gets reduced. The dimensional properties of the knitted fabric were significantly altered with the addition of elastane yarn in the fabric structure [2] as shown in Table 3.1.

While comparing the effect of different elastomeric yarn brands on the dimensional properties of the single jersey fabric, a study reported that the addition of elastane yarn in the structure increased the loop densities and weight of the fabric. The shorter loop lengths of the spandex yarn were preferred as it increases the stretch, but it also increases the density of course and wales of the fabric, and also the weight. While comparing the different brands of the elastane, it was noted that elastane with the largest tension and constant draw ratio provides the highest tension and tightness value [3]. A similar result was also reported by another researcher who analysed the effect of elastane in knit structure on dimensional characteristics of the fabrics. They noted that the addition of elastane increases the course per centimetre and wales per centimetre and results in a reduction in loop length. These changes in the fabric structure also increase the stitch density than the 100% cotton fabric [1] (Table 3.1).

Marmarali evaluated the effect of elastane yarn in the single jersey structure in every alternative feed (half plating) and also in every feed (full plating). The developed fabrics were dried on a flat surface for a week after the washing and dyeing process. The effect of elastane plating on the dimension stability of the fabrics was reported after every stage. The results showed that the drying and relaxing process reduced the course space in all fabrics and after dyeing, a slight increase was noticed. As the tightness factors and loop length maintained constant, researchers noted a very less reduction in course spacing after half plating compared to the 100% cotton single jersey fabric. However, there is no change in the full-plated structure as it remains tight already. Similarly, the increment in the elastane content significantly increased the weight and thickness of the fabric proportionally [4]. Other researchers analysed the dimensional characteristics of core-spun cotton/spandex fabric at different drying conditions. The course and wales density increased after the subsequent relaxation process in both cotton and cotton/spandex fabric. The study reported a higher increment in the wales and course density in the case of cotton/spandex than the cotton fabric. Upon the correlation analysis between course and wales per centimetres with loop length, they have noted higher dimensional stability of cotton/

TABLE 3.1

Effect of elastane on dimensional properties of the knitted fabric

Sample type	Structure	Wales/cm	Course/ cm	Loop shape factor (1/ cm)	Tightness factor (ne/mm)	Loop length
100%	Single jersey	14	21.35	1.6	2.1	3.30
Cotton	Single pique	12.15	15.5	1.4	1.7	2.99
fabric	Fleece	11.14	16.8	1.53	1.3	4.20
	1×1 rib	12	13.35	1.2	1.6	3.17
	Interlock	12.98	11.6	0.9	2.4	2.19
95% Cotton	Single jersey	16.9	29.45	1.8	1.7	2.98
and 5%	Single pique	15.37	20.35	1.4	2.2	2.45
elastane	Fleece	14.43	21.6	1.53	1.6	3.61
	1×1 rib	17.5	7.5	0.38	1.3	4.12
	Interlock	20.94	7.6	0.3	1.7	2.98

Source: [2] (Reprinted under creative commons license)

spandex fabric than the 100% cotton fabric. This is mainly associated with the quick recovering elastic nature of the spandex blended yarn, as it relaxes and reaches a stable condition than the cotton fabric [5].

While most of the studies reported on single jersey fabric, Chathura and Bok reported the rib structure. After different relaxation processes, the results revealed that under given experimental conditions, the cotton/spandex fabric was noted with more length and widthwise shrinkage than the cotton fabric after the 10th laundry cycle [6]. Both 100% cotton and cotton/spandex fabric showed higher shrinkage or dimensional change with loose structure than the structure with tight and medium tightness factors. The lower shrinkage associated with the tight fabric may be due to the completely packed highly dense structure that restricts the relaxation of the fabric. But in the case of cotton fabric, with respect to structural compactness, no difference was noted with each relaxation cycle. The researcher also evaluated the spirality values (specifically wale distortion) of the rib structure. Generally, a higher tightness factor offers lower spirality values, and for the balanced structures like 1×1 rib, it is always less. When the cotton and cotton/spandex fabrics were compared, a lower spirality angle was noted in the case of cotton/spandex fabric after 10 laundry cycles. Though both cotton and cotton/spandex fabric showed acceptable changes in spirality (less than 5°), a lower change was noted with cotton/spandex fabric [6]. Table 3.2 represents the test methods used by other researchers to measure the dimensional characteristics of the fabric.

The dimension stability of the elastane woven fabric was analysed with three different structures namely plain, twill, and satin. When dimensional changes were considered, a significant effect of elastane draw ratio was noted than the weave structure. They have reported an increasing contraction with an increase in the draw ratio.

TABLE 3.2

Test methods used for the measurement of dimensional characteristics

Fabric properties	Test methods/instruments used for the measurement	References
Loop length, wales per cm (wpc) and	Crimp tester and pick glass	[2, 5]
course per cm (cpc)	ISO 7211–2	[3]
	TS EN 12127:1999	[15]
Stitch density	*Stitch density* (S) = wpc $*$ cpc.	[2]
	TS EN 14971	[17]
Tightness factor	*Tightness factor* $(k) = \sqrt{Tex}/l$.	[2]
	TS 7128 EN ISO 5048:1998	[15]
Mass per unit area	ISO 33071	[1]
	TS 251	[4]
	ISO 3801	[3]
	TS 250 EN 1049-2:1996	[15]
	ASTM D3776/D3776M-09a	[9, 20]
	TS EN 12127	
Thickness	TS7128	[4]
	ISO 5084	[3]
	TS 7128 EN ISO 5084	[17]
Relaxation process	ASTM D 1284-76	[5]
Shrinkage	ISO 6330	[1]
Spirality	AATCC 179	[1]

Out of all the three structures evaluated, a higher shrinkage or dimensional change was noted with twill fabric based on drawing ratio. The researcher pointed that the higher retraction value of the spandex with a higher draw ratio was the main reason for the higher-dimensional change. Whereas the structural effects were considered a maximum effect noted with satin fabric followed by twill and plain weaves [7]. The effect of elastane (Lycra) proportion on the fabric dimensional properties was reported by Abdesselam et al. [8]. The effect of elastane draw ratio (core-spun yarn) and yarn linear density on the dimensional properties of the fabrics in the weft direction were also reported [9]. The findings of the study showed that an increase in both draft ratios of the elastane linear density, increased the dimensional change of the fabric. Out of these two parameters concerned, when the draw ratio was kept constant, the increase in the liner density alone had a significant effect on the contraction of the fabric. When the significance of the effect was considered, the influence of linear density on dimensional change was noted higher than the draw ratio. This was mainly attributed to the higher elastane proportion in the fabric [9]. It also noted that the inclusion of spandex in the cotton fabric significantly influences the loop shape factor of the cotton/spandex fabric compared to 100% cotton fabric. These changes significantly affect the dimensional properties of cotton/spandex fabric. However, in the case of polyester/spandex blend, no such relationship was reported by the researcher [10].

Other researchers evaluated the effect of dual-core elastane in the weft direction of the woven fabric and measured their impact on dimension stability. As reported in the previous studies, after washing, the woven fabric showed widthwise changes. However, there is no much change in widthwise reduction of the fabric, with the increase in elastane content. They reported an inverse relationship between the weft elastane density and shrinkage percentage. Similarly, the researcher also noted a linear relationship between the elastane density and fabric weight [11]. Though the elastane core-spun yarn used in the weft direction, a complete warp and weft shrinkage was reported by another study. In which the dimensional change of the fabric was evaluated after dry relaxation and after home laundry. They have reported a reduction in dimensional change when the elastane draw ratio was reduced. In summary, the addition of elastane in the knit or woven structure increases the tightness of the fabric and makes the fabric dimensionally stable after different drying and relaxation process. Whereas in the case of woven fabric more widthwise shrinkage was reported and in the knit fabric, an increase in course density was reported. In both structures, fabric weight increased significantly after the elastane inclusion. In the case of knit structure, the elastane addition modifies the loop shape and alters the loop shape factor. Further, when the elastane yarn is considered, either as core-spun or by a plating method, a higher elastane percentage aided a firmer structure. When the elastane is considered, yarn linear density caused a higher impact on the dimensional property of the fabric followed by the elastane draw ratio. The effect of elastane content on the dimensional properties of the knitted fabrics was summarized in Figure 3.1.

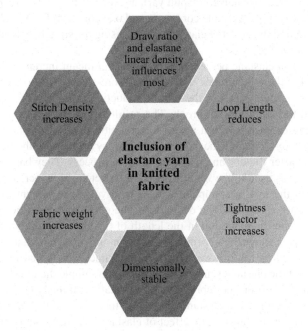

FIGURE 3.1   Effect of elastane content on the fabric's dimensional properties.

3.3   STRETCH AND RECOVERY PROPERTIES

Elastane has an extensibility range of 300%–500%. Elastane usage in fabrics increased as its inclusion in normal fabric phenomenally increased the extensibility of the fabric. The elastane-incorporated fabric generally possess low resistance against the force acting on it. It also possess a lower elastic modulus and a higher breaking elongation value than that of normal fabric. The common proportion in which the elastane is included in the fabric ranges from 2% to 8% [12]. When a limited load has been applied, the fabric with elastane will recover completely and at extensive application may recover partially or based on the time of a load applied. As we have already discussed, the inclusion of elastane in a lower percentage also affects the fabric dimensional property significantly. Hence, this section details the stretch and recovery properties of elastane-included textile material. To evaluate the effect of elastane content on the stretch and recovery property, researchers used plain-woven fabric with different elastane content. The researchers developed the fabric with a cotton core-spun yarn in fabric with different weft insertion rates from one cotton yarn and one elastane core-spun yarn, two elastane and one cotton, four elastane and one cotton, six elastane and one cotton, and completely all weft as elastane. On stretch and elastic recovery rate analysis, the study results showed that an increase in the elastane content in the weft of the woven fabric increases the fabric stretch level significantly ($P > 0.01$). The lower stretch value (6%) was noted with one elastane and one cotton fabric, whereas the 15% of extension was noted with elastane alone used fabric. Similarly, in line with these results, a higher recovery rate of 82% was noted with fabric that used elastane core-spun yarn at every weft. A lower recovery rate of 78% was noted with one elastane core-spun and one cotton weft inserted fabric. The study demonstrated the effect of elastane in increasing the stretch and recovery of the plain-woven fabric [13].

The increase in elastane percentage showed a significant increment in the breaking elongation of the woven fabric with elastane core-spun weft yarn. The researcher used different elastane ratios in the fabric production (4%, 5%, 7%, 9%, and 11%) and measured the elongation at break. The results showed up to a 75% increment in the breaking extension with a positive correlation with the elastane ratio in the fabric [14]. A study evaluated the stretch and recovery property of the core-spun weft elastane fabric after the dry and domestic laundry process. They have reported that as the elastane was used in the weft direction, the stretch value increased in the weft direction when the load was applied. However, a very low amount of stretch was noted in the warp direction. The decrease in the elastane draw ratio increases the stretch level due to the reduction in the length of the elastane core yarn. The home laundered samples showed no correlation between the draw ratio and elastic nature as like dry relaxed fabric. However, high stretch values were noted in the warp and weft direction of the elastane samples. The laundered sample did not show any effect of elastane draw ratio on the fabric stretch. On statistical analysis, the researchers found a significant impact of elastane draw ratio, the load, and the relaxation type on the stretch value of the fabric. The effect of elastane draw ratio and amount of load on the stretch property was greatly influenced by the type of relaxation process. Out of all the three parameters considered, the stretch level of the elastane fabric was majorly affected by the relaxation method. When the permanent deformations were analysed, the laundered fabrics showed a lower residual formation with a lower elastane draw ratio and load. Hence, the study suggested a lower values (3.07 elastane draw ratio and 25N load) for the textiles that have the aesthetic application [15].

Gorjanc and Bukosek evaluated the stretch and recovery properties of one way (weft alone elastane) and two way (both warp and weft) stretch fabrics. After the elastane addition, the stretch fabrics showed a lower elastic modulus, lower extension at yield for all the three fabrics. This lower yield is a representation of the lower elastic deformation of the fabrics. The researchers noted a higher breaking extension in the two-way stretch fabric than the one-way fabric. It is reported that at lower loads within the yield point, the extension does not influence the core elastane. However, loads beyond the yield stretched and caused transversal pressure in the stretch direction. The effect of long duration extension (one hour) on the elastic and viscoelastic behaviour of the fabric was studied. Two-way fabrics can stretch and recover more than the one-way fabrics used. The study reported the importance of elastane content in the fabric on the viscoelastic properties of the fabric even after longer duration stretching [16]. Nilgun Ozdil evaluated the twill woven fabric with different elastane percentages for their elastic stretch and recovery properties. The results of the study showed that the increase in the elastane content in the fabric (weft yarn) increases the maximum stretching and elastic recovery rate of the fabric. In the case of permanent stretching, the increment in elastane content significantly reduces it [18].

The effect of weft yarn density on stretch and permanent elongation of the fabric was evaluated with dual-core elastane yarn (weft) woven denim fabric. The results of the study reported a reduction in elasticity with an increase in the weft yarn density. The results are reported in Table 3.3, and it can be noted that the higher elasticity of 70% was noted with a weft density of 12 thread/cm, whereas in the case of weft density of 30 thread/cm, an elasticity of 20% was noted. As far as the stretchability is concerned, it was noted that the impact of weft yarn density was higher than that of the elastane ratio. Hence, the researchers advised to go with lower weft (elastane) density for the applications that require a higher elasticity. In the case of permanent elongation, a similar effect is reported by the researchers [11].

Payal Bansal et al. reported the effect of elastane linear density on the stretch and recovery property of the core-spun elastane yarn (inserted as weft) fabric. The researchers evaluated the immediate elastic recovery, delayed elastic recovery, and permanent set value of the fabric. An increment in the elastane yarn linear density increased the immediate recovery by representing the finer yarns which recover better and faster than the courser yarn. As elastane was used in the weft direction, the overall recovery was higher in the weft direction compared to the warp. When the delayed recovery and permanent set were concerned, similar results were reported. Courser yarns showed more restriction and lower elastic recovery than the finer elastic yarns. As for the direction was concerned, higher recovery and lower permanent set were noted in the weft direction. In the case of stretch percentage also a similar effect was noted concerning immediate, delayed recovery and permanent set [19]. A similar study also reported that the increase in weft density significantly reduced the breaking elongation, elasticity, and fabric growth value. In the case of core-spun yarn types, the double core yarns were found to be more elastic than the single

TABLE 3.3
Effect of weft density on the elasticity and permanent elongation of fabrics

Weft density (thread/cm)	Elasticity (%)	Permanent elongation (30 s) (%)	Permanent elongation (2 h) (%)	Elasticity/ permanent elongation (2 h)
12	74	12	8.8	8.41
14	62.8	10.2	8	7.85
16	52	9.6	6.8	7.65
18	44	7.6	5.4	8.45
20	41	7	4.8	8.54
22	33	5.4	3.8	8.68
24	28.2	5.4	3.4	8.29
26	26.4	5.6	3.6	7.33
28	20.6	4	2.4	8.58
30	20	4	2.4	8.33

Source: [11] (Reprinted with permission)

core-spun yarn [20]. To predict the long-term stretch properties of the elastane fabric, researchers evaluated the stretch property up to 25 washes. The results showed that the stretchability of the laundered fabric increased significantly from 15% (unwashed) to 30%–34% after five cycles. However, the increment in the stretchability remained constant till the 25 washes. This was mainly associated with the release of initial tension associated with the fabric till the fifth wash. Further, the results confirmed that the repeated laundry did not affect the stretch properties of the elastane woven fabrics [21].

In the case of knitted fabric, the effect of half and full plating on the stretch and recovery properties of the fabric was evaluated by Selin Hanife Eryuruk and Fatma Kalaoglu [22]. The results are provided in Table 3.4. It can be noted that the increase in the elastane content increases the elastic properties, as the plating changes from half to full plating. When the elastane linear density was considered, an increase in elastane linear density reduced the extension and residual extension of the fabric. Further, an increment in the elastane yarn count increased the widthwise extension of the knitted fabric. Overall, higher extensibility was noted with the full-plated knitted fabric than the half-plated knitted fabric. Researchers also reported that the amount of elastane and linear density increment increases the weight and density of the fabric. In the case of 40Ne (ground yarn) fabric, a higher elastic recovery was noted at the lower-elastane-linear-density fabrics compared to the higher-elastane linear-density fabrics. Whereas, in 30 Ne fabric, a higher extension was noted with the courser elastane yarn knitted fabric. However, they did not discuss the causes for the differences [22].

In summary, when the stretch and recovery properties are considered, whether it can be a knitted or woven fabric, the elastane percentage increases with increase in stretch percentage, breaking elongation, and recovery. A full-plated knitted fabric and two-way stretch fabric (woven) can possess higher extension than a half-plated

TABLE 3.4.
Elastic recovery properties of half and full plated knitted fabric with different ground yarn count

	Elastane plating	Elastane yarn Count (dtex)	Recovery at 30 min		Extension		Residual Extension	
			Widthwise %	Lengthwise %	Widthwise %	Lengthwise %	Widthwise %	Lengthwise %
A1	Half	22	77.7	91.8	121.5	58.50	22.3	8.2
A2	Half	33	80.6	91.0	111.1	61.90	19.4	9.0
A3	Half	44	85.0	88.8	106.7	75.85	15.0	11.2
B2	Full	33	85.7	87.9	135.5	126.50	13.3	12.1
B3	Full	44	90.1	88.0	116.9	143.90	9.9	12.0
C1	Half	22	57.4	90.2	184.4	88.50	42.6	9.8
C2	Half	33	62.0	90.4	169.6	93.0	38.0	9.0
C3	Half	44	61.0	91.6	183.3	98.20	39.0	8.4
D1	Full	22	81.4	90.4	170.1	133.5	18.6	9.6
D2	Full	33	70.4	82.2	198.9	165.2	29.6	17.8
D3	Full	44	74.6	80.4	179.5	180.5	25.4	19.6

Source: [22] (Reprinted with permission)

and one-way stretch fabric. A higher width or weftwise extension was noted in both woven and weft knitted fabric as the elastane included mainly widthwise. When the elastane linear density was considered, an increase in linear density significantly reduced the extensibility and recovery characteristics of the fabric. Similarly, a lower elastane draw ratio showed a higher extension due to a lower elastane core. However, research findings reported that higher elastane density (number of elastane yarn per square centimetre) in a fabric results reduction in fabric elongation. The type of relaxation method employed (dry or wet) has a significant impact on the elastic properties of the fabric. A wet relaxation method reduces the internal binding stress of the fabric and increases the extensibility of the fabric to a certain level. After those initial relaxation or increase in stretchability, the wet relaxation method did not influence the stretch and recovery properties of the elastane fabric till 25th wash. Various standards used to measure the stretch and recovery properties of the elastane fabrics are provided in Table 3.5.

3.4 BAGGING PROPERTIES OF ELASTANE FABRIC

The shape of the garments undergoes numerous deformations during the use. However, these changes are temporary due to the elastic nature of the structure and fibres used in the fabric. When the applied pressure is higher enough to make permanent changes in the structure, that is beyond the fabric's viscoelastic behaviour, which can cause irreversible deformations [23]. In body parts like the elbow, knees and hips impart a lot of stress on the fabric. A minimum of 20%–30% of the stretch level is required for the performance apparel. Whereas a stretch level of 25%–40% is essential for the garments that requires comfort [24]. Bagging is one of such permanent deformation of fabric or apparel, in a three-dimensional form and it deteriorates the appearance of the garment. A detailed summary of various definitions of bagging was reported by Vildan Sülar and Yasemin Seki [25] and the fundamental theory and mechanisms of bagging can be found elsewhere [26]. They have reported that the bagging problem occurs when multiaxial cyclic stress is applied to a textile, due to the function of fibre (fibre type, blending ratio, resilience, and viscoelastic properties), yarn (type, linear density, and twist), fabric (type, weight, thickness, and finishing), and mechanical properties (shear rigidity, formability, bending, compression and extensibility). This section of the chapter details various effects of elastane yarn on the bagging properties of both woven and knitted fabrics.

Using an instrument that depicts the movement of the knee and elbow, Nilgün Özdil evaluated the bagging properties of denim fabric made of different percentages of elastane ratios. Their finding showed that an increase in the elastane content in fabric significantly reduces the permanent bagging effect in the fabric, in the meantime, the elastic bagging increases proportionally [18]. Bagging properties of plain-woven fabric with different weft elastane densities were analysed using Instron tensile strength tester with clamp modifications [27]. The results of the research showed that fabric extensibility and sample diameter used in the testing significantly influences the fabric bagging properties. An increment in the elastic yarn density significantly decreases the fabric residual bagging height and bagging fatigue. The bagging height reports the fabric deformation (maximum) at load, compared to the higher at no load and the bagging fatigue (a phenomenon that relates to the elastic and viscoelastic nature of the fabric) shows the loss of fabric energy on the repeated cyclic load. When the density of the elastane yarn was reduced, the residual bagging height increased due to the lower extensibility of the fabric. The researchers reported that an increased fabric extensibility showed a significant reduction in fabric tensile elastic modulus and so it reduces the restriction towards fabric bagging deformation. A similar trend was also noted with fabric fatigue. As the increment in fabric extensibility reduces the inter yarn frictional forces, the restriction on the deformation reduces and so does the fabric fatigue. The results are found to be statistically significant [28]. Baghaei et al. measured the bagging behaviour of elastane woven fabric under tensile fatigue cyclic loads as per another report [29]. The research evaluated the effect of elastane draw ratio and core-spun elastane yarn twist on the bagging properties of the woven fabric. It was noted that a higher draw ratio increases the elasticity of the fabric and it also possesses higher strength to withstand the cyclic load. Whereas in the case of yarn twists, a lower twist level supported higher elasticity of the fabric. Fabric produced with higher twist yarns showed a higher fatigue value after cyclic load due to its lower elasticity. The permanent fatigue significantly increased till 1,000 cycles; however, further increment till 2,000 cycles did not show much impact on the bagging fatigue of the fabric [30].

The bagging property of the single jersey knitted fabric with different elastane densities was evaluated by Seval Uyanik and Kubra Hatice Kaynak [17]. The study analysed the bagging property of the knitted fabric in two methods by applying repeated cyclic load (100 and 200 cycles) and also by changing the bursting height at different heights (50%–80% of burst heights). The results of the study evaluated the first extension before fatigue loading and the last extension after the end of the 100th and 200th cycles of loading. The results denoted that only the single jersey fabric without elastane content showed an increased extension value after the cyclic load application. In the case of other fabrics with elastane, all the fabrics showed a lower extension irrespective of the elastane content. The difference between the first and last (after cyclic load application) extension value is noted higher after the 100th cycle, but in the case of the 200th cycle, the difference was noted lower due to the permanent deformation of the fabric. These findings confirmed the elastane usage can reduce the fabric bagging significantly than the fabric without elastane. The study also evaluated the bagging property with respect to the bursting height, namely 50%, 60%, 70%, and 80% of burst height repeated for 100 cycles and the bagging properties reported. The results showed that except for the fabric without elastane, all the fabric showed a lower value in the last extension results at all bursting height percentages. In all the fabric, both the first and last extension values increased for every bursting height mentioned. Despite these changes, the full elastane sample did not show any changes after all the treatment. It delivered the same value in all extension levels. This result confirms that sample without elastane, 1×1 elastane, and 2×1 elastane samples increases the bagging value with increment in the load or extension. In the case of 2×2 elastane fabric, no bagging was observed after all the tests performed [17].

The residual bagging height in the knitted fabric with and without elastane yarn was predicted with fuzzy theory. The study compared the practical results and evaluated the theoretical model that predicts them. Their findings reported that in knit structures, the residual bagging height decreases with the use of elastane yarn. An increment in the elastane percentage significantly reduces the bagging. The research also reported that the addition of elastane increases the bending rigidity that makes the fabric stiffer [31]. When compared to the single jersey fabric, the 1×1 rib knitted structures showed a lower residual bagging height due to their lower internal stress [32]. The effect of sample size and yarn structure on the residual bagging height was measured and reported that sample size significantly influences the residual bagging height. Similarly, the yarn structure has its potential impact. A higher elastic yarn showed a reduction in residual bagging height than the normal nonelastic yarn. Further, the research also reported that the anisotropy was also important to characterize the behaviour of knitted fabrics. When the elastic and nonelastic fabrics were compared, the bagging share tendencies showed different internal stress levels of different fabrics. The residual bagging height increased with a reduction in the gauge value. The shear stiffness of the fabric differs based on slipperiness at loop intersections. They have concluded that the addition of elastane yarn in fabric structure significantly reduces residual bagging height. Further, careful control of input parameters like blend type, gauge length, and yarn structure along with elastic filaments has significant control on the bagging behaviour of the knitted fabric [33].

From the results analysed, it can be noted that repeated cyclic load and exceeded pressure beyond the elastic nature will cause the bagging appearance in fabrics. In the case of both knit and woven fabric, the increment in elastane content increases the elasticity and reduces the bagging property. The inclusion of elastane yarn reduces inter yarn friction and reduces residual bagging height significantly. A higher draw ratio was suitable for increasing the elastic nature and so it reduces the bagging property of the fabric. In the case of yarn twists in the core-spun yarn, a lower twist factor reduced the bagging property due to their higher elongation. When the different ratios of elastane were noted, all yarns with elastane showed a lower bagging ability to repeated cyclic load for 200 cycles. Other than elastane, yarn structure, knitting machine gauge, and anisotropy of the factor were also noted to have a significant effect on the bagging behaviour and residual bagging height of the fabric. Table 3.5 represents different test methods adopted by previous researchers to analyse the bagging behaviours of fabrics.

3.5 DYNAMIC ELASTIC PROPERTIES

Dynamic elastic behaviour of elastane fabric denotes the dynamic work recovery and stress at the specific extension of fabric [34]. This property of textile fabric is important to analyse the instant response of the fabric to body movement. Nonelastic textile materials generally possess a viscoelastic deformation with a hysteresis loop, a perfect elastic recovery is required to get a complete dynamic work recovery. Garment fit, slip, and fabric stretch are the important parameters that influence skin movement. Where, fit of the garments is the ratio between the body size and garment size. Slip is associated with the coefficient of friction, and stretch accommodates the skin strain. When the fabric stretch is considered, it directly influences the pressure comfort of the garment, which is based on garment stretch and recovery properties [35].

TABLE 3.5

Test methods used for the stretch and recovery, bagging, and dynamic elastic recovery property measurement of elastane fabric

Fabric properties	Test methods/instruments used for the measurement	References
Stretch and recovery	ASTM D2594-04	[13]
	ASTM D1682	[14]
	BS EN 14704-1:2005 (CRE type test device to extend	[15]
	the specimen to a specified load principle)	[16]
	DIN 53835	[18]
	TS 6071 and ASTM 3107	[11, 19–21]
	ASTM D3107	[22]
	BS 4958-1992.	
Bagging property	DIN 53860	[18]
	Instron tensile stretch tester with some clamp	[28]
	modification as per [27]	[30]
	Zhang et al.'s Instrument	[17]
	[29]	[31]
	TS EN ISO 13938-2 (Bursting strength method)	
	Pressure applied using steel balls	
Dynamic elastic recovery properties	ASTM D 4964–96 method (CRE principle)	[34, 38–41]

The property is called dynamic work recovery, which helps us in understanding the power gain of the wearer of the elastic garment while a maximum stretch extension was applied. Here the term work recovery must be clearly understood as it differs from elastic recovery. The elastic recovery is the ratio between the recoverable strain and total strain, whereas the elastic work recovery is the ratio of recovered elastic energy and total elastic recovery [34]. Though previous researchers used Kawabata Evaluation System (KES), for the measurement of dynamic work recovery using tensile resilience (RT%) methods under a constant rate of loading, the extension level that can be measured was limited to 5%–15% [36]. These amounts of stretch may not be enough to predict the actual body movement in a situation like sports applications, in which approximately 10%–50% of the extension is required. Hence, Senthilkumar et al. evaluated the dynamic elastic behaviour of the cotton and cotton spandex knitted fabric at the different extension levels. The elastic hysteresis value of the fabric was initially evaluated at different extension levels such as 20%, 30%, 40%, and 50% as shown in Figures 3.2 and 3.3 [37].

The results reported an increment in hysteresis along with the extension increment from 20% to 50% for both cotton and cotton/spandex fabric. Elastane-incorporated fabric showed a lower stress value than the cotton fabric in both wale and course direction as reported in Figures 3.2 and 3.3. The lower extension value of cotton/spandex fabric was mainly attributed to the lateral compression of yarn in elastane-incorporated structure. Hence, applied energies were used for the extension

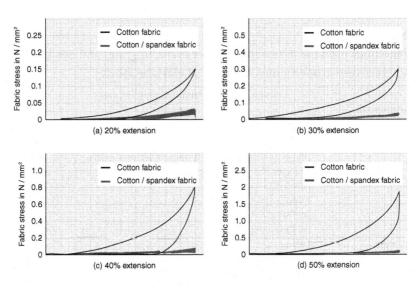

FIGURE 3.2   Elastic hysteresis of cotton and cotton/spandex fabrics walewise direction [37] (Reprinted under Creative Commons licence).

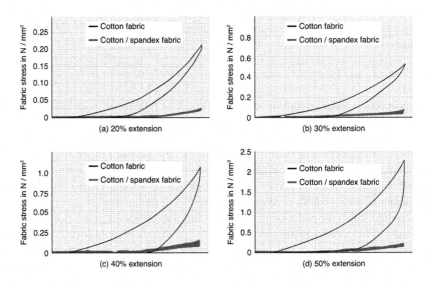

FIGURE 3.3   Elastic hysteresis of cotton and cotton/spandex fabrics coursewise direction [37] (Reprinted under Creative Commons licence).

of compressed loops than the fabric expansion, and so caused minimum stress on the wearer for all levels of extension applied [37]. With these values, the researcher evaluated the dynamic work recovery value using the following formula.

$$\text{Dynamic work recovery} = \frac{\text{Area under unloading curve}}{\text{Area under loading curve}} \times 100$$

The dynamic work recovery value of the cotton and cotton/spandex fabric in both wale and coursewise direction was provided in Table 3.6 [37]. The results showed that the dynamic work recovery value of cotton fabric was noted lower than the elastane fabric. The cotton/spandex fabric showed a 20% higher dynamic work recovery value on wales direction and a 15% on the course-wise [34, 37]. In the case of cotton fabric, a reduction in dynamic work recovery value was noted with an increase in extension from 20% to 50%. This was the case in both course and wale wise direction. Whereas, in elastane fabric, an initial increase in dynamic work recovery value noted up to 30% extension and further it reduces with 40% and 50% in both course and wales direction. One of the other important factors that should be considered while garment application is stress level at maximum extension. The analysis of stress level research reported that an exponential increment in the cotton fabric during the extension on both directions, from 20–50% (1.75 N/mm^2 to 2.3 N/mm^2). In the case of cotton/spandex fabric, a lower stress level was reported below 0.5 N/mm^2 in both directions for up to 50% extension [34, 37]. These results imply a cotton/spandex fabric imparts very low stress on the wearer at extensions and makes the wearer feel comfort than cotton fabric.

Zhi-Cai Yu et al. evaluated the dynamic elastic behaviour of warp knitted elastane fabric made up of various filament yarns. The researcher measured the dynamic elastic recovery of the fabric in walewise direction with 20%–50% extension levels. The results reported a higher extension in the wale direction than the course direction of the fabric. Though no much difference was noted in the elastic behaviour, an increment in dynamic elastic recovery was noted between 20% and 30%, and a slight reduction was noted at 30%–50% of extension level. The difference is mainly related to the overlapped loop extension in the elastane knitted warp fabrics. Whereas, at higher

TABLE 3.6
Dynamic work recovery value of cotton and cotton/spandex knitted fabric

Fabric specification	20% Extension	30% Extension	40% Extension	50% Extension
Cotton/spandex fabric (waleswise direction)	68.34	74.45	64.24	60.35
Cotton fabric (waleswise direction)	58.52	54.15	48.36	43.68
Cotton/Spandex fabric (coursewise direction)	65.34	73.88	71.34	68.45
Cotton fabric (coursewise direction)	60.52	60.15	56.49	52.07

Source: (Reprinted under creative commons license)

extension states, the elastane yarn extends completely and results in the loop slipping inside the fabric. Hence, at 50% of extension, a lower dynamic elastic recovery was noted [38]. Researchers also compared the effect of elastane core-spun yarn knitted fabric and elastane plated knitted fabric on the dynamic elastic behaviour. Upon stretch values of 20%–50%, the results showed that both the knit fabrics (with elastane core-spun yarn and plated yarn) had a similar effect for the extension levels. At the extension of 20%–30%, an increase in dynamic elastic behaviour was noted followed by a reduction at 30%–50%. Out of the two fabrics used, elastane-plated knitted fabric showed a higher dynamic elastic recovery value (23% on wales direction and 2% in course direction) than the elastane core-spun yarn fabric. When the fabric stress level was considered, a higher stress level was noted with elastane-plated knitted fabric than the core-spun fabric on both the directions [39]. A 25% and 24% higher stress level in the elastane-plated knitted fabric was noted in wales and course direction respectively [37].

The effect of elastane yarn input tension, loop length, and elastane linear density on the dynamic elastic recovery of the fabric was evaluated as per the standards reported in Table 3.5. The results reported that increase in elastane yarn loop length from 0.85 mm to 1.1 mm first increases and decreases the dynamic elastic recovery. A maximum dynamic elastic recovery was noted for the elastane loop length of 0.97 mm both in wales and coursewise direction. There was a 44% higher recovery rate noted in the wales wise and a 32% increment in the course direction. When the linear density of the elastane yarn increased from 20–40 denier, a fabric with higher linear density showed nearly 32% and 30% increment in the dynamic elastic recovery respectively in the wales and course directions. An increment in the elastane linear density increases the geometrical properties of the fabric and so increases area density. An increased elastane content and higher weight increase the residual energy and influence the dynamic elastic recovery. The increase in cotton yarn loop length (from 2.5 mm to 2.9 mm) initially reduces the recovery percentage and then increases. An overall, the study reported an insignificant effect of cotton yarn loop length on dynamic elastic recovery property [37, 40]. Effects of repeated laundry on the dynamic elastic recovery were also reported by Senthilkumar and Anbumani [41]. The study result showed that cotton/elastane fabric showed higher dynamic elastic recovery in both course and wales direction. An increase in the laundry cycle did not affect the dynamic elastic recovery of cotton fabric. As far as the cotton/elastane fabric was considered, a slight increase was noted with the increase in the wash cycle in both directions. However, after the 10th wash to the 20th wash, a reduction in dynamic elastic recovery was noted with the cotton/elastane fabric. Though differences were noted in the reading, statistical analysis results confirmed the lack of significant difference among the washing stage for both the fabric tested. These findings confirmed that the laundry process did not influence the dynamic elastic recovery property of selected fabrics [41]. To summarize the findings, it is clear that the addition of elastane content in the fabric increases the dynamic elastic recovery of the fabric in both the wale and course direction. Further, the performance of plated knitted fabric was noted higher than the elastane core-spun yarn knitted fabric. When the pressure at maximum extension was measured it was noted lesser in the elastane fabric than normal cotton fabric. The research also showed the insignificant effect of the laundry process. Though both weft and warp knitted fabric showed similar results, no research report was found on the dynamic elastic recovery properties of woven fabric.

3.6 PHYSICAL PROPERTIES OF ELASTANE FABRIC

As the addition of elastane in the fabric has a significant impact on the dimensional properties of the fabric, their influence on the physical properties is also important. Hence, this section details the general impact of the elastane yarn addition on the physical properties of both woven and knitted fabric.

3.6.1 TENSILE PROPERTIES/BURSTING STRENGTH

The tensile strength of the fabric represents the maximum load a fabric can withstand during the application. It helps us to understand the breaking stress at which the fabric failures. Woven fabric with different elastane core-spun yarn ratios in the weft (by varying the number of elastane per one cotton yarn) was evaluated for the tensile properties. The results showed that an increment in elastane content reduced the tensile strength of the fabric. Out of all the samples evaluated, a higher tensile strength value was noted for the one cotton one elastane fabric, and lowest strength noted for all weft elastane fabric [13]. Tensile properties of polyester/elastane knitted fabrics were measured at different elastane percentages (12%, 14%, and 16%) by other researchers [42]. The results showed the highest Young's Modulus, breaking load, and lowest breaking extension of 100% polyester fabric than the polyester/elastane knitted fabric. Further increment in the elastane yarn percentage increases the breaking extension and load and Young's modulus reduces. Bursting strength measures the biaxial tensile strength of the knitted textiles with the equal course and wale densities. In the case of polyester/elastane knitted fabric, the densities of the yarn were not equal on the sides. The elastane content in the fabric creates a skewed loop in the structure while applying the bursting load. Hence, the force acting on the blended fabric may not be perfectly biaxial. The results showed that 100% polyester fabric showed a higher bursting strength value than the polyester/elastane fabric. However, an increasing trend in bursting strength was noted with an increment in the elastane content of the fabric.

For a fabric with equal wale and course densities, the fabric jamming will occur in both directions at the same point and a fabric failure is expected and it will be a circular hole (after bursting). When the jamming in one direction occurs fast or earlier than the other direction. It is expected to have a different hole/deformation shape during the bursting test. In the case of polyester a perfectly circular hole was noted, whereas, in the case of polyester/elastane fabric, the wales reached earlier jamming and bursting occurred in different shapes as elastane content increases as illustrated in Figure 3.4. Elastane full-plated knitted fabric had a higher bursting strength value than the half-plated knitted fabric due to their higher elastane content. Concerning the elastane yarn linear density, an increase in elastane linear density increases the bursting strength of the fabric in both elastane half-plated and full-plated knitted fabric. The result showed a strong correlation between the elastane amount and the bursting strength of the knitted fabric [42]. A similar result was also reported by

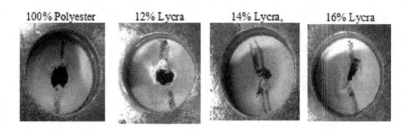

FIGURE 3.4 Fabric failure shapes of fabric with different elastane (Lycra) percentage (Reprinted under Creative Commons licence).

another researcher who compared the 100% cotton single jersey knitted fabric with elastane in every feed. The results showed a significant increment in the bursting strength from 124.1 KN/m^2 (cotton) to 146.3 KN/ m^2 [1].

Effect of relaxation and elastane draw ratio on the bursting strength of the bi-stretch woven fabric was reported by Kaynak [43]. As the true strength of the bi-stretch elastane fabric can be evaluated by the bursting strength method rather than the tensile strength, the researcher adopted this method. The results showed a significant reduction in the bursting strength of laundered textiles than the dry relaxed textiles. Further to add, there was no impact of elastane draw ratio on the bursting strength of laundered textile, whereas, in the case of dry relaxed textile, a decrease in the elastane draw ratio increased the bursting strength of the fabric. The same study reported a higher bursting height value of the fabric with a lower elastane ratio. A lower ratio offers more length of yarn in the loop and so it increases the breaking extension of the fabric compared to fabric with a higher elastane draw ratio [43]. The warp and weft knitted fabric with different elastane ratios in the fabric (2%, 3.5%, 5%) were analysed and reported for their bursting strength properties. In the case of both the warp and weft knitted fabric, an increase in elastane percentage increased the bursting strength value. Particularly in weft knitted fabric, the values analysed between different structures and reported a higher value with interlock followed by rib and single jersey fabric. All the structures responded similarly to the increment in elastane ratio [44].

Bilal Qadir et al. analysed the effect of elastane draw ratio and elastane yarn linear density on the tensile property of the woven fabric made of elastane core-spun weft yarn. The results of the study reported a negative correlation between elastane percentage in the fabric and the tensile strength. An increase in the elastane percentage reduced the tensile strength of the fabric. The authors reported two reasons for the effect namely, due to the higher breaking extension and lower tenacity of the elastane fibre, Secondly, the increase in elastane draw ratio reduces the elastane content in the yarn and so increase the inter-fibre friction and reduces overall yarn strength. When the tensile strength value of fabric made from a different denier of elastane yarn was compared, a lower tensile strength was noted with fabric produced in a higher denier of elastane. The study also compared the elastane percentage in the fabric and reported that irrespective of the elastane draw ratio and yarn linear density, the percentage of elastane in fabric decides the strength of the fabric [45]. Impact of weave type and elastane draw ratio of core-spun elastane yarn was reported by

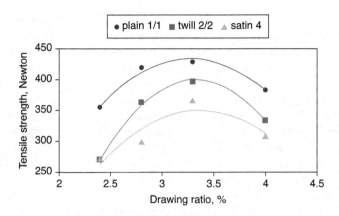

FIGURE 3.5 Effect of spandex filament drawing ratio on the tensile strength of cotton/spandex woven fabric with different weave structures (Reprinted with permission from [7]).

Alsaid Ahmed Almetwally and Mourad [7]. The study reported a nonlinear relationship between the elastane draw ratio and tensile strength in all selected structures. As the draw ratio increased from 2.4 to 3.3, an increment in the tensile strength was noted but the further increment of draw ratio to 4, reduced the tensile strength of the fabric. Statistical analysis revealed that the elastane draw ratio influences the fabric tensile strength to the level of 26%. Researchers supported this nonlinear relationship with a phenomenon called elastane fibres stress-induced crystallization. At a higher draw ratio, due to the higher stress, the molecules aligned straight and forms crystals due to the formation of hydrogen bonds. This increase in the crystalline nature reduces tensile strength [7]. In the case of weave type, a higher strength was noted with plain-woven fabric followed by twill and satin. Weave type accounted for a major influence on the tensile strength of the fabric up to 36%. The effect of draw ratio and weave type on tensile strength is provided in Figure 3.5 [7]. Among the different types of elastane yarn compared (full plating, core-spun, and dual core-spun), a higher bursting strength was noted with 100% cotton single jersey knitted fabric. A higher breaking extension was noted with a dual core-spun and core-spun elastane yarn fabric than the full-plated knitted fabric [48]. Similarly, when different brands of elastane yarns were evaluated, no significant difference was noted among the elastane brands [3].

3.6.2 TEARING STRENGTH

Tearing strength test is mainly performed for the woven fabrics to test their resistance against the tearing force. Woven fabric with different elastane weft densities showed a reduction in tear strength with the increment in the elastane content. A fabric with lower elastane content was noted with higher tear strength, whereas a higher elastane content showed a lower tearing strength. The researchers reported that reduction in elastane content developed a loose structure that allows grouping of yarn in structure and create higher resistance than the tighter elastane rich fabric [13]. Other researchers who compared the effect of elastane yarn linear density and elastane draw ratio on woven fabric reported a positive correlation between elastane content and tearing strength. The increase in the elastane percentage either by increasing the filament count or by decreasing the elastane draw ratio increases the tearing strength, whereas in the case of tensile properties, the correlation between them was negative. The difference is mainly due to the test method used. When the tearing force was applied to the fabric, the internal yarns slips and group together at the point of load and support each other. As the higher elastane content offers more slippage, a higher tearing strength was noted. They also mentioned that the draw ratio of elastane yarn highly influences the tearing strength. As the draw ratio increases, the stretchability of the elastane yarn was reduced and so it reduces the tearing strength of the fabric [45]. However, these findings were in contradiction with the previous results reported by other researchers [13].

While evaluating the effect of weft density on the denim fabric performance, Osman Gökhan Ertaş used the elastane containing dual core-spun yarn as weft. While increasing the weft density from 12 to 30 yarns per centimetre, a reduction in tearing strength was noted in both the warp and weft direction of the fabric. A tear strength reduction of 42% and 20% was noted from weft density 12 to 30 in the weft and warp direction respectively [11]. In a different study, the researcher used different weave structure effects and elastane draw ratio effects on the tear strength of the fabric. Statistical analysis results reported that 37% of tear strength-related properties were influenced by the elastane draw ratio. An increment in the elastane draw ratio reduces the tear strength value in all the fabric structures up to 11%–18%. This is associated with the fabric firmness increment with the elastane ratio increment. In the case of fabric structure, the satin fabric was noted with higher strength and plain fabric with lower tear resistance value was noted. The effect of draw ratio and fabric structure on tear strength is provided in Figure 3.6 [7].

3.6.3 Air permeability

Air permeability of the textile material is one of the important parameters as far as wearer comfort is concerned. As we include elastane in fabric to increase the fit and comfort, it is also necessary to analyse the impact of elastane on the air permeability

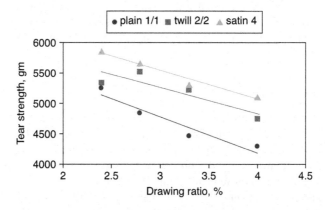

FIGURE 3.6 Effect of spandex filament drawing ratio on tear strength of cotton/spandex woven fabric with different weave structures. (Reprinted with permission from [7]).

of the textile. Research works reported that an increase in elastane density (yarns per centimetre) in the weft direction significantly reduces the air permeability value of the textile materials. A higher permeability value was noted with one elastane and one cotton alternative pick fabric than the complete elastane weft and six elastane and one cotton yarn fabric. The increment in the elastane content shrinks the fabric in the weft direction and hence the fabric compactness increases. A subsequent reduction in inter yarn spaces of the compact fabric leads to the lower air permeability values of the woven fabric with elastane than the normal fabric [13]. Other researchers measured the effect of elastane linear density, ground yarn count change, and elastane half and full plating on the air permeability of the elastane knitted fabric [22]. It was noted that elastane full-plated fabric showed lower air permeability than the half-plated knitted fabric due to the higher elastane density and compact structure. A significant reduction was noted in the half-plated fabric than the full-plated fabric. When the elastane linear density changed from 22 to 44 dtex, a reduction in air permeability was noted in both full and half-plated fabrics. Fabric with courser elastane yarn provides higher air permeability values [22]. Other researchers also reported a significant reduction in the air permeability value of the elastane full-plated knitted fabric than the cotton fabric [1, 55].

The air permeability value of dry relaxed and laundered bi-stretch elastane woven fabric showed a significant reduction in the laundered fabric than the dry relaxed fabric. This change was attributed to increased structural compactness and set value of the laundry relaxed fabric. When the elastane ratio was concerned in laundered textiles, due to its compact nature, no effect of elastane draw ratio was noted on the air permeability of the elastane fabric. However, in the case of dry relaxed fabric, a reduction in elastane draw ratio reduces the air permeability of the bi-stretch woven fabric [43]. Research on woven twill structures measured the effect of elastane dual core-spun yarn weft density and elastane draw ratio on the air permeability value of different twill fabrics. The results reported a reduction in the air permeability value with the increase of elastane draw ratio and weft yarn sett value of the fabric. Compared to 2/2 twill, a lower air permeability was noted with 3/1 twill fabric as

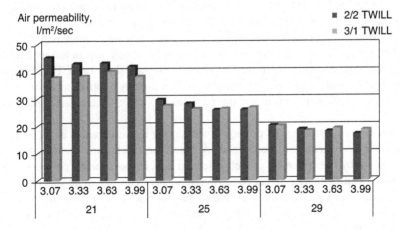

FIGURE 3.7 Air permeability results of 3/1 and 2/2 twill fabric samples with different weft density and elastane draw ratio [46] (Reprinted under Creative Commons licence).

shown in Figure 3.7. Statistical analysis showed a significant influence of weft sett value but the insignificant effect of elastane draw ratio and weave type of the fabrics tested [46].

Various types of knit structures were produced with both cotton and cotton/elastane blended yarn (5% elastane) and compared for their air permeability values. The results showed that irrespective of the structure of the knitted fabric, all elastane-contained fabric showed a statistically significant reduction in the air permeability. Researchers reported that the addition of elastane increases the thickness, stiffness, and tightness of the fabric and so the reduction in the air permeability was noted with the elastane fabric compared to cotton knit structure [47]. Other researchers reported a higher influence (85%) of elastane draw ratio on the air permeability of the fabric than the fabric structure (26%). An increase in elastane draw ratio reduces the air permeability of the fabric up to 30% in different structure. At a higher ratio, the increase in elastane percentage increases the fabric contraction, thickness, and fabric compactness. As a result, the air permeability value reduces with the changes in elastane draw ratio. As far as weave structure was concerned a higher air permeability was noted with plain fabric followed by twill and satin. Though satin fabric had a higher float of threads, due to their higher fabric thickness, structural compactness of the satin fabric than the other fabric [7]. When the different types of elastane yarns were compared, a higher reduction in air permeability was noted with dual-core and core-spun elastane yarn. Compared to 100% cotton fabric, the difference in air restriction was lower with full-plated single jersey knitted fabric [48]. A similar effect was noted when the researcher compared different brands of elastane knitted fabrics. There was no difference among the elastane brands with respect to the air permeability value [3].

As the physical properties were concerned, an increase in elastane percentage generally reduce the tensile strength of the woven fabric and bursting strength of the knitted fabric. In the case of elastane yarn linear density, an increment in elastane yarn count reduces the strength of the fabric. But few researchers also reported an increase in the bursting strength of the fabric. Though the majority of the works fall in this category few researchers also reported contradictory results on the tensile and bursting strength with respect to the elastane properties, while few researchers reported a reduction in tearing strength while increasing the elastane percentage. However, a reduction in tearing strength was noted as common with all researchers when there is an increase in elastane draw ratio. These findings indicates the requirement of future studies in this area by controlling several other fabric parameters which influence the physical properties like tensile and tearing strength.

There was no difference of opinion noted among the researchers as far as the air permeability properties are concerned. An increase in elastane content, either by draw ratio or by linear density or weft insertion rate, reduces the air permeability. In the case of knitted fabric, higher reduction in air permeability value was noted with the full-plated knit fabric compared to half-plated. Similarly, among different elastane fabrics, a higher air permeability reduction was noted in the fabric with core-spun elastane yarn than the plated fabric. Various test methods used in the physical properties evaluation by different researchers are provided in Table 3.7.

TABLE 3.7
Physical and mechanical properties of elastane fabric and test methods reported by other researchers

Fabric properties	Test methods/instruments used for the measurement	References
Tensile properties	ASTM D1682	[7, 13]
	ASTM D5035	[45]
Air permeability	ASTM D737	[7, 13, 47]
	TS 391 EN ISO 9237	[22, 43, 46]
	ISO 9237	[1, 3]
	EN ISO 9237	[48]
Tearing test	ASTM D1424	[7, 11, 13, 45]
Bursting strength	ASTM D 6797-02	[42]
	TS 393 EN ISO 13938-1	[22, 3]
	ISO 13938-21999	[1, 43]
	ASTM D3787-2001	[48]
Crease recovery	AATCC 66	[7]
	IS 4681-1968	[49]
	ASTM D-1295-67	[51]
Stiffness and skewness and bow measurements	ASTM D 4032-08	[9]
	ASTM D 3882-08	[20]
Stiffness	ASTM D4032	
Abrasion resistance	ASTM D4966-98	[50]
	TSE ISO 12,947	[20]
Pilling resistance	ASTM D4970	[50, 51]

3.7 MECHANICAL PROPERTIES

Mechanical properties like stiffness, bending rigidity, crease recovery, drape were discussed in this section. The details of various test methods used for the analysis of mechanical properties by different researchers are provided in Table 3.7.

3.7.1 CREASE RECOVERY

The addition of elastane filament in the knitted and woven fabric structure influences the crease recovery properties. A study evaluated the effect of fabric structure and elastane draw ratio on fabric properties. In the case of woven fabric, an increase in elastane draw ratio had a significant increment in the crease recovery angle. A higher increment was noted with satin structure than the plain and twill fabric used in their study. An increment in the elastane draw ratio increases the stretchability of the elastane in the fabric and so a higher recovery rate was observed. Statistical analysis revealed that a total of 55% contribution was noted from the elastane draw ratio. When the weave structure was considered, a higher recovery was noted with the twill structure due to its loose structure than the plain and satin used. [7]. Other researchers did not find any changes in the crease recovery of the twill woven fabric produced from different elastane percentages. The researcher used different elastane deniers to increase the elastane content from 1% to 3% [49]. The increase in elastane percentage from 0 to 8% increased the crease recovery of the single jersey knitted fabric up to 25%. An increase in elastane percentage increases the elasticity of the fabric and so increases the crease recovery [51].

3.7.2 DRAPE COEFFICIENT

Drape coefficient determines the fabric's ability to fall or drape over an object due to the gravitational force. A lower drape coefficient value shows a better drape ability of the textiles. The addition of elastane content to the fabric had a significant impact on the drape coefficient value. The study reported a considerable increase in the drape coefficient value of the twill woven fabric while increasing the elastane yarn linear density. Due to the increase in linear density, the elastane proportion increases along with fabric compactness and weight, which ultimately helps in increasing the drape coefficient value [49].

3.7.3 STIFFNESS, SKEW, AND BOWNESS

The increment in the weft density increases the stiffness of the fabric. An increment in the elastane percentage showed an increase in the fabric stiffness. While comparing the stiffness of elastane core-spun and dual core-spun yarn, the latter had higher stiffness [20]. The presence of dual elastane yarn increases the fabric contraction and so increases the stiffness. In stretch denim fabric, skewness, bowness, and stiffness were noted as one of the quality issue. The effect of elastane yarn linear density and draw ratio on the stiffness, skew, and bowness properties of the fabric were analysed by Muhammad Bilal Qadir et al. [9]. An increment in the elastane linear density and elastane draw ratio in the weft direction had a significant effect on the fabric stiffness. Out of the two parameters evaluated, the effect of the elastane draw ratio was noted higher with a Pearson correlation coefficient of 0.94 than the elastane linear density (0.562). An increase in elastane draw ratio increases fabric mass per square metre and contraction along with an increase in stiffness [9]. Due to the construction parameters, fibre, and yarn characteristics, sometimes the warp and weft yarns in the fabric may not be at right angle and displaced angularly. This was known as Skewness. When the elastane was added into the woven structure as weft yarn, it increases the crimp and contraction of the fabric in the weft direction and affects the skewness. When the elastane density and draw ratio was considered, a higher influence of the elastane draw ratio was noted than the elastane linear density. A similar effect was also noted in the fabric bow percentage. The influence of the elastane draw ratio was higher than the elastane yarn linear density. An increase in contraction of the fabric in addition to elastane percentage was reported as an important factor [9].

3.7.4 ABRASION CHARACTERISTICS

Effect of elastane content on the single jersey, interlock, rib, single pique, and fleece fabric on the abrasion resistance were studied. The results showed that the addition of elastane content (5%) significantly improved the abrasion resistance value of all structures. Out of all the structures tested, single jersey fabric showed higher resistance than the other structure followed by single pique, 1×1 rib, and interlock and fleece knitted fabrics. The order was the same in the case of 100% cotton fabric due to the impact of structural changes. Hence, researchers concluded that the elastane content significantly improves the abrasion resistance of the fabrics. The effect of fabric type was noted as dominant [50]. Other researchers compared the effect of single core and dual core elastane yarn on the abrasion resistance of one-way stretch fabric and reported no significant changes after 25,000 cycles in both warp and weft directions [20].

3.7.5 PILLING RESISTANCE

The pilling resistance of the cotton elastane knitted fabric increased while compared to the 100% cotton fabric. As researcher compared the addition of 5% elastane in the yarn and measured the effect on the pilling resistance on single jersey, rib, interlock, single pique, and fleece structures. Among the structures analysed single jersey fabric showed the highest resistance against the pilling with and without elastane content [50]. Other researchers measured the effect of elastane percentage on the single jersey knitted fabric. When the elastane percentage increased from 0 to 8%, the pilling resistance of the fabric increases significantly. The addition of elastane content increases the resistance by increasing the structural compactness of the fabric [51].

3.8 LOW-STRESS MECHANICAL PROPERTIES

Das and Chakraborty measured the effect of core-spun elastane yarn in weft on the low-stress mechanical properties of the fabric. Researchers used a statistical method to understand the interactive effect of different process parameters like elastane stretch, the proportion of elastane content, and the yarn twist multiplier. The results of the research reported a significant effect of elastane on the low-stress mechanical properties of the woven fabric [52]. An increment in elastane stretch percentage reduced the breaking load. No influence of elastane proportion on low-stress breaking elongation was reported. An increment in the elastane percentage reduces the shear rigidity and shear hysteresis of the fabric. With respect to the shear hysteresis, an increment was initially noted with an increase in elastane proportion and it drops. As far as the bending rigidity of the fabric is considered, an increment in the elastane stretch reduces it. However, the elastane proportion did not have any influence on the bending rigidity. A similar result was also reported for the shear hysteresis by the authors. When the linearity of the compression was considered, an increase in elastane stretch decreased it, and an increase in elastane proportion showed an increment in the linearity of the compression. However, no relationship was noted between the compression resilience of the fabric and elastane proportion [52].

Other researchers measured the hand value of the single jersey knitted fabric developed using elastane full-plated, elastane core yarn, and dual-core elastane yarn. Using KES system, researchers evaluated both the primary hand value and the total

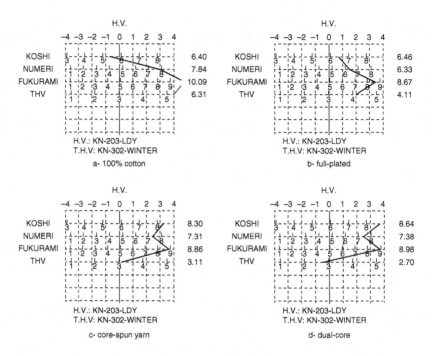

FIGURE 3.8 Primary hand value and total hand value of different elastane fabric [48] (Reprinted under Creative Commons licence).

hand value of the fabric. Core yarn contains cotton to elastane ratio of 86 to 14 and the dual-core yarn contains polyester and elastane along with cotton. The results showed that the addition of elastane content reduced the total hand value of the fabric significantly. Compared to the 100% cotton fabric, full-plated single jersey knit fabric showed a 17% reduction followed by core-spun and dual core-spun yarn with a 38% and 45% reduction in total hand value respectively. The results of the primary and total hand value of KES analysis are provided in Figure 3.8 [48].

3.9 MOISTURE PROPERTIES OF ELASTANE FABRICS

Moisture properties of the textile are very essential concerning the comfort aspects of the fabric. Though several research works were performed on the dimensional and physical properties of the elastane knitted fabrics, the moisture-related properties were not measured completely. A very few studies analysed different moisture handling characteristics and their findings are summarized in this section.

3.9.1 WICKING BEHAVIOUR

Wicking characteristics of textile material describe their moisture uptake ability through capillary action. This parameter is one of the important comfort aspects of the textile as it helps in absorbing and transferring of sweat from the skin. In a comparative analysis of the full-plated and half-plated knit structure of single jersey and

1×1 rib, researchers measure the fabric dimensional and moisture characteristics. The study reported a significant level of difference in both longitudinal wicking and transfer wicking. In the case of transfer wicking, the highest wicking value was noted for half-plated fabric than the 100% cotton and full-plated fabric. An increase in the elastane content reduced the transfer wicking of both the single jersey and 1×1 rib structures. The study inferred that the addition of elastane content in the fabric reduced the absorbing material content (cotton) of the fabric and hence the transfer wicking value showed less in the full-plated structures of single jersey and rib structures. The second potential cause that was reported was an increment in the elastane content in a full-plated fabric that produced a compact structure than half-plated fabric. The researcher also reported that higher wicking of half-plated fabric than 100% cotton (both single jersey and rib) fabric was due to the increase in structural compactness, which provides more absorbing capacity initially, and hence an increment noted but after a point the value reduces with the increment of elastane (full-plated). In the case of longitudinal wicking, an increase in the elastane content significantly increased the wicking. The addition of elastane content increased the compactness of the fabric and reduced the pore size in structure. As per capillary principle, smaller pores first fill and moves upward faster to larger holes due to the capillary pressure. Hence, higher wicking was noted both in the course and wales direction. When the fabric structures were compared, higher longitudinal wicking was noted in the plain fabric than the rib due to its structural variations [53].

Another study compared the transfer wicking characteristics of 100% cotton and cotton/elastane fabric after different finishing treatments. Among the different finishing treatments, cotton fabrics showed higher wicking after softening treatment compared to cotton/elastane fabric. In contrast, the resin finished fabrics showed lower wicking values in both type of fabrics [54]. Manshahia and Das evaluated the filament shape, loop shape factor and elastane linear density on the moisture characteristics of polyester/elastane plated knitted fabric. The results reported that an increase in elastane linear density significantly influences moisture absorption and in-plane wicking characteristics. An increase in elastane linear density reduces the wicking and absorption capacity of the elastane plated polyester fabric. The increment in linear density increases the fabric compactness and reduces the moisture characteristics [55]. Among the different structures of knitted fabric with 100% cotton and cotton/elastane yarns, a significant impact of the structure was noted on vertical wicking. However, no such relationship was found with the elastane composition [59].

3.9.2 Water Vapour Permeability

The ability of the textile material to allow moisture or water vapour through its structure can be measured using this method. Nuray Kizildag et al. measured the effect of elastane plating on the water vapour permeability of the 1×1 rib and single jersey fabrics. Though results showed a reduction in the rib fabric, in both the fabrics, elastane plating reduced the water vapour permeability significantly. The researcher related the permeability reduction with the increased compactness of the elastane plated fabric [53]. In a study that compared the fabric with different types of elastane insertions like plating, core-spun, and dual core-spun on water vapour transmission.

The results showed that a full-pated knitted fabric showed a higher vapour permeability than core-spun and dual core-spun fabric. It was reported that higher structural compactness and lower porosity of the elastane core-spun and dual core-spun fabrics than full-plated and 100% cotton fabric [48]. While comparing the half-plated and full-plated knitted fabrics, the study reported a reduction in the water vapour permeability of the full-plated fabric than the half-plated fabric. The study also considered the effect of elastane yarn linear density on water vapour permeability. The results were similar to air permeability results. An increase in the linear density reduced the permeability value of the fabric. Increased structural compactness and reduction in pore size due to the increase in elastane percentage were noted. This is the main reason for vapour permeability reduction in the fabrics [55]. While comparing single jersey fabric with different elastane percentages (5.5%, 6%, and 6.5%) and different elastane insertion types, researchers did not show any significant effect on water vapour permeability of the fabric [58]. Water vapour permeability value of five different structures namely single jersey, rib, interlock, single pique, and fleece fabric was evaluated with 100% cotton yarn and cotton/elastane (95%/5%) blended yarn. The results showed that out of all the fabric single pique fabric showed higher water vapour permeability value with and without elastane. Based on the structural openness and compactness, the water vapour permeability differed among the fabric. When the elastane was introduced, there was a significant reduction in the water vapour permeability value of the fabric irrespective of the structures. The findings showed that the higher structural compactness and increased weight of the fabric were due to the elastane addition [59]. A 20% increment in the water vapour resistance value of woven fabric with weft elastane yarns was reported by Dunja Sajn Gorjanc et al. Studies measured the water vapour permeability of cotton/elastane (98/2%) yarn after different finishing treatments. They measured the water vapour permeability of the fabric after rigid finishing, resin finishing, bleaching, and softening. The results were in contradiction with the previous works [53] and reported a higher vapour permeability in the cotton/elastane fabric than 100% cotton fabric. These changes may be related to the finishing treatment that denim fabric has undergone. Hence, the results of the study cannot be compared with the other studies of water vapour permeability [54].

3.9.3 DRYING BEHAVIOUR

The drying time of the fabric had a very specific role in the comfort characteristics of the textile as it determines the wetness and dryness of the skin. Compared to the cotton fabric, cotton/elastane fabrics showed a faster drying rate due to the elastane content and lower absorption rate. This is due to the addition of hydrophobic elastane, the amount of cotton fibres in the yarn cross-section or area of the fabric reduces and so the drying rate of the fabric increases. The same trend was noted with fabric even after a different finishing process [54].

3.9.4 MOISTURE MANAGEMENT INDICES

Moisture management indices correlate the various fabric characteristics and overall moisture management capacity (OMMC) value of any textiles. As elastane fabrics

TABLE 3.8

Moisture management characteristics of different elastane fabrics

MMT parameters	100% Cotton yarn	Fully plated yarn	Core-spun yarn	Dual core-spun yarn
Wetting time top (s)	15.99	15.28	6.44	8.71
Wetting time bottom (s)	6.31	9.86	17.49	16.68
Top absorption rate (%/s)	39.29	33.41	38.88	42.29
Bottom absorption rate (%/s)	83.60	87.60	31.32	34.41
Top max wetted radius (mm)	17	14	16	13.75
Bottom wetted radium (mm)	18	16	17	16.25
Top spreading speed (mm/s)	0.99	1.12	1.11	0.86
Bottom spreading speed (mm/s)	1.43	1.27	0.81	0.72
Accumulative one way transport capacity	382.89	276.67	-63.77	30.10
Overall moisture management capacity	0.655	0.597	0.086	0.160

Source: [48] (Reprinted with creative commons license)

are commonly used in tight wear like athletic and sportswear, it is necessary to analyse the effect of elastane content on fabric moisture properties. A study measured the moisture management indices of full-plated, core-spun, dual core-spun yarn, and 100% cotton fabrics. Compared to the 100% cotton and full-plated fabrics, a poor one-way transport capacity (OWTC) was noted with the core-spun and dual core-spun elastane fabric. This might be due to the reduction in the fabric porosity and increased compactness due to a high elastane percentage. This represents the poor sweat management characteristics of core-spun and dual-core elastane fabrics compared to plated fabric. A higher OMMC value reported a higher moisture handling ability of the textile. The detailed results of moisture studies are provided in Table 3.8. The study showed a higher OMMC value for the cotton and elastane full-plated fabric. The higher OMMC value of the fabric (Closer to 1) represents better moisture handling. However, elastane core-spun and dual core-spun showed poor moisture characteristics with low OMMC value [48].

In a similar study, other researchers reported that increment in the elastane linear density significantly reduces the wetting time of the fabric. Along with loop length reduction and higher elastane density, an increment in fabric wetting time was noted. Similarly, the absorption rate of the fabric gets reduced while increasing the elastane linear density. These changes are associated with higher elastane content and higher structural compactness of the fabric. The research reported a significant reduction in OWTC of the fabric with an increase in elastane linear density. Similar to the previous research, this study also mentioned that increment in the elastane percentage in the fabric deteriorates the comfort and moisture management properties of the fabric significantly [55].

3.10   THERMAL PROPERTIES

Thermal characteristics of textile primarily deal with the heat transfer characteristics of textile material between skin and environment. As the thermal properties highly influence the comfort characteristics, it is necessary to understand the effect of elastane inclusion on the thermal characteristics of the elastane fabric. A study reported the different thermal characteristics like thermal conductivity, and thermal absorptivity of the elastane knitted fabric. The thermal conductivity coefficient (λ) represents the amount of heat, which passes from 1 m^2 area of material through the distance 1 m within 1 s and the temperature difference 1K [57]. Research work compared the effect of linear density on the thermal conductivity value of both full and half-plated fabrics. The results of the study reported that an increase in the elastane linear density increased the thermal conductivity of the fabric. However, an increase in fiber fineness from 20 D to 40 D did not show any significant difference. But the fineness increment from 20 D to 60 D has shown significance. When the structure of the fabric was compared, an increase in elastane percentage increased the thermal conductivity. A higher value was noted with full-plated knitted fabric over the half and 100% cotton fabric. The thermal conductivity of the fabric was mainly attributed to the fabric openness and the entrapped air in the structure. A loose structure will lead to higher air entrapment and reduces the thermal conductivity. When the elastane was added, a tighter and compact structure reduced the entrapped air in the structure and hence increases the thermal conductivity [56]. While analysing the effect of elastane linear density, Manshahia and Das reported that the fabric with courser elastane yarn was capable of more conduction of heat than the finer yarn. The researcher reported that the increased fabric density and thickness were the reason for lower air entrapment and higher conductivity [55].

Amany Khalil et al. studied the effect of elastane percentage of the elastane core-spun knitted single jersey fabric on thermal conductivity and thermal resistance. The results showed that compared to elastane core-spun knit structure, full-plated knit fabric showed a higher thermal conductivity and lower thermal resistance. The changes might be attributed to the tighter structure of the fabric as reported by previous researchers [58]. But on the contrary, other researchers reported an increment in the thermal resistance with the increment in the elastane percentage of the fabric [48, 59]. An increment of 20% in the thermal resistance value of the woven fabric with elastane in the weft direction was noted with 100% cotton fabric [60].

The researchers also evaluated the thermal absorptivity of the fabric. It is a contact temperature measurement that depicts the warm-cool feeling. A textile material with a lower thermal absorptivity value shows a warmer feeling of the material. The results of the study reported that an increment in the elastane linear density and elastane percentage increases the thermal absorptivity value. It can be seen that higher

TABLE 3.9

Fabric moisture properties and test methods reported by other researchers

Fabric properties	Test methods/instruments used for the measurement	References
Wicking	DIN 53924	[53]
	ASTM D-5802	[55]
	AATCC 197	[59]
Water vapour permeability	TS EN ISO 11092	[56]
	ISO 11092	[58]
	ISO 811-201	[59]
	ASTM E96-00.	[53, 54, 60]
Moisture management test	ISO 9073-8	[48]
	AATCC 195: 2011	[55]
Thermal properties	Alambeta Tester	[48, 56, 58]
	ASTM D 5470	[59]
	ALMENO 2590	[60]

elastane content increases the thermal conductivity value and so the fabric remains cool. Compared to the 20 denier yarn and elastane half-plated knit structure, a 60 denier fabric and full-plated knit fabric showed a higher absorptivity and cool feeling to the wearer [56]. In another research, the effect of elastane core-spun, dual core-spun and full-plated fabric was compared. The results showed that a full-plated fabric showed a 20% increase in the absorptivity values. Whereas the elastane core-spun, dual core-spun fabric showed a 27% increment in the absorptivity values than 100% cotton fabric. This represents that the increase of elastane percentage increases the cool feeling of the wearer [48]. The elastane core-spun and full-plated knitted fabric was compared with three different elastane percentages for their thermal absorptivity. The value decrease with an increase in elastane percentage. Compared to the core-spun yarn a full-plated fabric showed the lowest value and warm feeling. However, the effect was statistically insignificant [58]. The details of different test methods used in the measurement of moisture properties are provided in Table 3.9.

The overall results showed few contradictory results in wicking and thermal properties. All the research findings reported that an increment in the elastane percentage reduces the water vapour permeability of the fabric significantly. In the case of wicking, few researchers supported the increment in the wicking ability of the fabric, while others reported the negative effect with the addition of elastane yarn in the fabric. A similar trend was also noted with thermal conductivity and thermal resistance of the fabric with elastane content. Compared to the core-spun and dual core-spun yarns, higher wicking, higher thermal conductivity, and higher vapour permeability were noted with the full-plated knitted fabric. In the case of moisture management properties also a higher OMMC and OWTC values were noted for full-plated knitted fabric than the core-spun yarn knitted fabrics. An overall increase in elastane percentage either by proportion or by elastane linear density significantly reduces the OWTC and OMMC value of the fabric.

3.11 SUMMARY

The chapter represents the effect of elastane yarn addition in fabric structure on the various properties of fabrics. The review results showed that the incorporation of elastane yarn showed a serious change in the dimensional properties of the fabric. It was also evident that changes in dimensional properties like increased thickness, compactness, and weight of the fabric showed a higher influence on the other physical, mechanical, and moisture-related properties. In general, the inclusion of a higher elastane percentage in the fabric reduces the moisture and comfort-related properties and increases the tensile, stretch, and dynamic elastic recovery properties. However, in the case of thermal resistance and wicking, few researchers reported contradicting results. Similarly, very few studies were noted in the handle and moisture management-related properties. These findings suggest the requirements of future studies to standardize the elastane effect on fabric properties.

REFERENCES

1. Abu Yousuf Mohammad Anwarul Azim, Kazi Sowrov, Mashud Ahmed, Rakib Ul Hasan, H.M., Md. Abdullah Al Faruque. (2014). Effect of elastane on single jersey knit fabric properties – Physical & dimensional properties. *International Journal of Textile Science*, 3(1), 12–16. DOI: 10.5923/j.textile.20140301.03.
2. Dereje Berihun Sitotaw. (2018). Dimensional characteristics of knitted fabrics made from 100% cotton and cotton/elastane yarns. *Journal of Engineering*, 2018, Article ID 8784692, 1–9.
3. Serkan Tezel, & Yasemin Kavusturan. (2008). Experimental investigation of effects of spandex brand and tightness factor on dimensional and physical properties of cotton/spandex single jersey fabrics. *Textile Research Journal*, 78(11), 966–976. DOI: 10.1177/0040517507087685
4. Marmarali, B.A. (2003). Dimensional and physical properties of cotton/spandex single jersey fabrics. *Textile Research Journal*, 73(1), 11–14.
5. Chathura, N.H., & Bok, C.K. (2008). Dimensional stability of core spun cotton/spandex single jersey fabrics under relaxation. *Textile Research Journal*, 78(3), 209–216.
6. Chathura N.H., & Bok, C.K. (2007). Dimensional characteristics of core spun cotton-spandex core spun cottonspandex rib knitted fabrics in laundering. *International Journal of Clothing Science and Technology*, 19(1), 43.
7. Alsaid Ahmed Almetwally, & Mourad, M.M. (2014). Effects of spandex drawing ratio and weave structure on the physical properties of cotton/spandex woven fabrics. *The Journal of The Textile Institute*, 105(3), 235–245. http://dx.doi.org/10.1080/00405000.2013.835092
8. Abdessalem, S.B., Abdelkader,Y.B., Mokhtar, S., & Elmarzougui, S. (2009). influence of elastane consumption on plated plain knitted fabric characteristics. *Journal of Engineered Fibers and Fabrics*, 4(4), 30–35. https://doi.org/10.1177/155892500900400411
9. Muhammad Bilal Qadir, Zulfiqar Ali, Ali Afzal, Muhammad Irfan, Tanveer Hussain, Mumtaz Hassan Malik, Muhammad Waqas Iqbal, Amir Shahzad, Adnan Ahmad, & Zubair Khaliq. (2020). Effect of elastane parameters on the dimensional and mechanical properties of stretchable denim fabrics. *Clothing and Textiles Research Journal*, 40(2), 1–15. https://doi.org/10.1177/0887302X20968812
10. Gokarneshan, N., & Thangamani, K. (2010). An investigation into the properties of cotton/spandex and polyester/spandex knitted fabrics. *The Journal of The Textile Institute*, 101(2), 182–186, https://doi.org/10.1080/00405000802332032

11. Osman Gökhan Ertaş, Belkıs Zervent Ünal, & Nihat Çelik. (2016). Analyzing the effect of the elastane-containing dual-core weft yarn density on the denim fabric performance properties. *The Journal of The Textile Institute*, 107(1), 116–126. https://doi.org/10.1080/00405000.2015.1016319

12. Rupp, J., & Bohringher, A. (1999). Elastanhaltige garne und stoffe. *International Textile Bulletin*, 35(1), 10–30.

13. Mourad, M.M., Elshakankery, M.H., & Alsaid, A.A. (2012). Physical and stretch properties of woven cotton fabrics containing different rates of spandex. *American Journal of Science*, 8, 567–572.

14. Al ansary, M.A.R. (2011). Effect of spandex ratio on the properties of woven fabrics made of cotton/spandex spun yarns. *American Journal of Science,* 12, 63–67.

15. Hatice Kubra Kaynak. (2017). Optimization of stretch and recovery properties of woven stretch fabric. *Textile Research Journal*, 87(5), 582–592.

16. Gorjanc, D.S., & Bukosek, V. (2008). The behaviour of fabric with elastane yarn during stretching. *Fibres and Textiles in Eastern Europe*, 68, 63–68.

17. Seval Uyanik, & Kubra Hatice Kaynak. (2019). Strength, fatigue and bagging properties of plated plain knitted fabrics containing different rates of elastane. *International Journal of Clothing Science and Technology*, 31(6), 741–754.

18. Nilgün Özdil. (2008). Stretch and bagging properties of denim fabrics containing different rates of elastane. *Fibres and Textiles in Eastern Europe*, 16(1), 63–67.

19. Payal Bansal, Subhankar Maity, & Sujit Kumar Sinha. (2020). Elastic recovery and performance of denim fabric prepared by cotton/lycra core spun yarns. *Journal of Natural Fibers*, 17(8), 1184–1198. https://doi.org/10.1080/15440478.2018.1558151

20. Tuba Bedez Ute. (2019). Analysis of mechanical and dimensional properties of the denim fabrics produced with double-core and core-spun weft yarns with different weft densities. *The Journal of The Textile Institute*, 110(2), 179–185. https://doi.org/10.1080/00405000.2018.1470451

21. Kan, C.W., & Yuen, C.W.M. (2009). Evaluation of the performance of stretch denim fabric under the effect of repeated home laundering processes. *International Journal of Fashion Design, Technology and Education*, 2(2–3), 71–79. https://doi.org/10.1080/17543260903302329

22. Selin Hanife Eryuruk Fatma Kalaoglu. (2016). Analysis of the performance properties of knitted fabrics containing elastane. *International Journal of Clothing Science and Technology*, 28(4), 463–479. http://dx.doi.org/10.1108/IJCST-10-2015-0120

23. Nyoni, A.B., & Brook, D. (2010). The effect of cyclic loading on the wicking performance of Nylon 6.6 yarns and woven fabrics used for outdoor performance clothing. *Textile Research Journal*, 80, 720–725.

24. Kirk, J.W., & İbrahim, S.M. (1966). Fundamental relationship of fabric extensibility to anthropometric requirements and garment performance. *Textile Research Journal*, 36, 37–47.

25. Vildan Sülar, & Yasemin Seki. (2018): A review on fabric bagging: The concept and measurement methods. *The Journal of The Textile Institute*, 109(4), 466–484.

26. Şengöz, N.G. (2004). Bagging in textiles. *Textile Progress*, 36(1), 1–64. http://dx.doi.org/10.1533/jotp.36.1.1.59475.

27. Abooei, M., & Shaikhzadeh Najar, S. (2007). The effect of sample diameter and test speed on bagging behaviour of worsted fabrics. Paper presented at 9th Asian Textile Conference, Taiwan. http://dspace.lib.fcu.edu.tw/bitstream/2377/3885/1/ce05atc902007000014.pdf.

28. Shaikhzadeh Najar Saeed, Momeny Zahra, & Etrati Seyed Mohammad. (2018). Bagging behaviour of extensibile shirt fabrics, annals of the university of oradea fascicle of textiles, leatherwork. XIX(1), 95–100. https://doaj.org/article/631abdf2d7534c50b26ec5b86293ad78

29. Baghaei, B., Shanbeha, M., & Ghareaghaji, A.A. (December 2010) Effect of tensile fatigue cyclic loads on bagging deformation of elastic woven fabrics. *Indian Journal of Fibre and Textile Research*, 35, 298–302.
30. Zhang, X., Li, Y., Yeung, K.W., & Yao, M. (1999). Fabric bagging: Part I: Subjective perception and psychophysical mechanism. *Textile Research Journal*, 69, 511.
31. Boubaker Jaouachi. (2013). Evaluation of the residual bagging height using the regression technique and fuzzy theory. *Fibres and Textiles in Eastern Europe*, 21(100): 92–98.
32. Demiroz, A., & Dias, T. (2000). A study of the graphical representation of plain-knitted structures, Part II: Experimental studies and computer generation of plain-knitted, structures. *Journal of Textile Institute*, 91(4), 481–492.
33. Gazzah, M., & Jaouachi, B. (2014). Evolution of residual bagging height along knitted fabric lengths. *Research Journal of Textile and Apparel*, 18(4), 70–75. http://dx.doi.org/10.1108/RJTA-18-04-2014-B008.
34. Senthilkumar, M., & Anbumani, N. (2014). Dynamic elastic behavior of cotton and cotton/spandex knitted fabrics. *Journal of Engineered Fibers and Fabrics*, 9(1), 93–100.
35. Kirk, Jr., & Ibrahim, S.M. (1966). Fundamental relationship of fabric extensibility to anthropometric requirements and garment performance. *Textile Research Journal*, 36(1), 37–47.
36. Kawabata, S., Postle, R., & Masako, N. (1982). Objective specification of fabric quality, mechanical properties and performance", Proceedings of the Japan – Australia Joint Symposium, Kyoto, 1–29. https://www.worldcat.org/title/objective-specification-of-fabric-quality-mechanical-properties-and-performance/oclc/11115773
37. Senthilkumar, M. (2014). Dynamics of elastic knitted fabrics for tight fit sportswear, Doctoral thesis, Anna University, Chennai, India. http://hdl.handle.net/10603/15503
38. Zhi-Cai Yu, Jian-Fei Zhang, Ching Wen Lou, Hua-Ling He, An-Pang Chen, & Jia-Horng Lin. (2015). Wicking behavior and dynamic elastic recovery properties of multifunction elastic warp-knitted fabrics. *Textile Research Journal*, 85(14), 1486–1496. DOI: 10.1177/0040517514566105
39. Senthilkumar, M., & Anbumani, N. (2011). Dynamics of elastic knitted fabrics for sports wear. *Journal of Textile Institute*, 41, 13–24.
40. Senthilkumar, M., & Anbumani, N. (2012). Effect of spandex input tension, spandex linear density and cotton yarn loop length on dynamic elastic behaviour of cotton/spandex knitted fabrics. *Journal of Textile, Apparel Technology and Management*, 7(4), 1–16.
41. Senthilkumar, M., & Anbumani, N. (2012). Effect of laundering on dynamic elastic behaviour of cotton and cotton/spandex knitted fabrics. *Journal of Textile, Apparel Technology and Management*, 7(4), 1–10.
42. Shaimaa Youssef El-Tarfawy. (2016). Failure behavior of polyester/lycra single jersey knitted fabric. *American Journal of Science*, 12(5), 52–58. doi:10.7537/marsjas12051606
43. Kaynak, H.K. (2017). Effects of elastane draw ratio of core-spun yarn on air permeability and bursting strength of bi-stretch woven fabrics. *Journal of Textile Science and Engineering*, 7, 323. doi: 10.4172/2165–8064.1000323
44. Abd El-Hady, R.A.M. (2016). The influence of elastane ratio on bursting strength property of knitted fabrics. *International Journal of Advance Research in Science and Engineering*, 5(2), 1–10.
45. Bilal Qadir, Tanveer Hussain, & Mumtaz Malik. (2014). Effect of elastane denier and draft ratio of core-spun cotton weft yarns on the mechanical properties of woven fabrics. *Journal of Engineered Fibers and Fabrics*, 9(1), 23–31.
46. Çelik, H.I., & Kaynak, H.K. (2017). An investigation on the effect of elastane draw ratio on air permeability of denim bi-stretch denim fabrics. 17th World Textile Conference AUTEX 2017 – Textiles – Shaping the Future. IOP Conference Series: Materials Science and Engineering, 254. doi:10.1088/1757–899X/254/8/082007

47. Dereje Berihun Sitotaw. (2020). Air permeability and stiffness of knitted fabrics made from 100% cotton and cotton/elastane yarns. *Ethiopian Journal of Textile and Apparel*, 1(2), 2–9.
48. Amany Khalil, Abdelmonem Fouda, Pavla Těšinová, & Ahmed S. Eldeeb. (2021). Comprehensive assessment of the properties of cotton single Jersey knitted fabrics produced from different lycra states. *AUTEX Research Journal*, 21(1), 71–78. http://dx.doi.org/10.2478/aut-2020-0020
49. Sunny Pannu, Meenakshi Ahirwar, Rishi Jamdigni, & Behera, B.K. (2020). Effect of spandex denier of weft core spun yarn on properties of finished stretch woven fabric. *International Journal of Engineering Technologies and Management Research*, 7(7), 21–32. https://doi.org/10.29121/ijetmr.v7.i7.2020.720
50. Dereje Berihun Sitotaw, & Rotich K. Gedion. (2020). Performance characteristics of knitted fabrics made from 100% cotton and cotton/elastane blended yarns. *Journal of Textile Apparel Technology and Management*, 11(2), 1–13.
51. Haji, M.M.A. (2013). Physical and mechanical properties of cotton/spandex fabrics. *Pakistan Textile Journal*, 1, 52–55.
52. Das, A., & Chakraborty, R. (2013). Studies on elastane-cotton core-spun stretch yarns and fabrics: Part II – Fabric low-stress mechanical characteristics. *Indian Journal of Fibre and Textile Research*, 38, 340–348.
53. Nuray Kizildag, Nuray Ucar, & Belgin Gorgun. (2015). Analysis of some comfort and structural properties of cotton/spandex plain and 1×1 rib knitted fabrics. *The Journal of The Textile Institute*, 107(5), 606–613. http://dx.doi.org/10.1080/00405000.2015.1054143.
54. Selin Hanife Eryuruk. (2020). The effects of elastane and finishing properties on wicking, drying and water vapour permeability properties of denim fabrics. *International Journal of Clothing Science and Technology*, 32(2), 208–217. DOI 10.1108/IJCST-01-2019-0003
55. Manshahia, M., & Das, A. (2014). Thermophysiological comfort characteristics of plated knitted fabrics. *The Journal of The Textile Institute*, 105(5), 509–519. http://dx.doi.org/10.1080/00405000.2013.826419
56. Gözde Ertekin, Nida Oğlakcioğlu, & Arzu Marmarali. (2018). Strength and comfort characteristics of cotton/elastane knitted fabrics. *Tekstil ve Mühendis*, 25(110), 146–153. https://doi.org/10.7216/1300759920182511010
57. Hes, L. (2007). Thermal comfort properties of textile fabrics in wet state. Proceedings of XIth International Izmir Textile and Apparel Symposium, 26–29 October, Turkey.
58. Amany Khalil, Pavla Tešinová, & Abdelhamid R.R. Aboalasaad. (2019). Thermal comfort properties of single jersey knitted fabric produced at different Lycra states. 4th International conference on natural fibers (ICNF-2019), July, 1–3, Porto, Portugal.
59. Dereje Berihun Sitotaw, & Desalegn Atalie Wellelaw. (2020). Thermal, permeability and some dimensional characteristics of knitted fabrics made from 100% cotton and cotton/elastane yarns. *Journal of Textile Apparel Technology and Management*, 11(4), 1–13.
60. Dunja Sajn Gorjanc, Krste Dimitrovski, & Mateja Bizjak. (2012). Thermal and water vapor resistance of the elastic and conventional cotton fabrics. *Textile Research Journal*, 82(14), 1498–1506. DOI: 10.1177/0040517512445337

4 Elastane in garment fit and comfort

4.1 INTRODUCTION

Fit and comfort are two very important parameters concerning clothing. Comfort is generally defined as a pleasant state of physiological, psychological, and physical harmony between a human being and their environment [1]. A garment with sufficient room for body movement will result in comfortable wear. To achieve this comfortable feel, the garment must be free from unwanted wrinkles, gather, and should have a proper fit to the body. Hence, fit is defined as the way a garment conforms to the body [2]. Slater classified the garment comfort in three different categories as follows [1]:

1. **Physiological comfort**: Details regarding the body temperature, tactile sensation, neural responses, blood pressure, etc.
2. **Psychological comfort**: Details the mind's ability to feel satisfactorily without external help. Comfort is mainly affected by fear, stress, pleasure, or pain.
3. **Physical comfort**: Effect of external environment on the equilibrium of other two components of comfort.

A garment that fits perfectly will enhance the aesthetic look of the wearer and so the wearer feels psychologically confident. Further, the garment falls smoothly on different body contours without clinging, pulling, bending, or twisting. In contrast, the poorly fitted garments make the wearer feels so insecure and makes them feel total discomfort [3]. On the other hand, fit-related issues often create dissatisfaction, make shopping a burden for the consumers, and also make the customers emotionally depressed. Stamper et al. reported that the parameters like grain, ease, line, balance, and set are the elements of garment fit [2]. Understanding these parameters will enhance the knowledge of the fit and related issues. The importance of the fit elements on the fit characteristics of the apparel was provided in the following section.

4.2 FACTORS INFLUENCING GARMENT FIT

The following elements contribute positively to the final appearance of the garment. Proper use of these elements in a garment style or design enhances the fit of the garment.

i. **Garment ease**: The measurement difference between the apparel parts and actual body parts is known as garment ease. Ease was identified as one of the vital parameters that influence the garment's fit and comfort [4]. To

DOI: 10.1201/9780429094804-4

maintain comfort, the construction or assembling methods contribute much to the ease along with the design ease. The design's ease is crucial as it decides the fit of the garment in the preparatory stage itself.

  ii. **Fabric grain**: The term grain represents the direction of warp yarn in the fabric. Generally, the fabric grain decides the draping ability of the fabric. The alignment of grain in each cut component decides the functional and aesthetic qualities of the apparel. The perfect arrangement of weaves, the right angle between warp, produces higher drape and balanced grain and aids good fit to the wearer [5].

 iii. **Line**: Lines in a garment creates a visual feel of garment fit. Lines can be of garment designs or sometimes may be of grain, seams, darts tucks, or pleats. A vertical line makes the silhouette looks taller. The correct flow of line in the body contour without any baggy look or wrinkles represents a good fit.

  iv. **Garment set**: The formation of wrinkles in a garment represents the poor fitting of the garment. The smooth final appearance of a garment shows a set. In other words, the absence of wrinkles in the garment shows a good garment set. Formations of wrinkle, based on the direction or locations, indicate the poor fit [6].

   v. **Balance**: The balance of a garment is obtained from its identical or symmetrical correlation between the garment and body. A perfect balance in a garment can be achieved by its proper design. A balanced garment must-have garment parts of similar scales.

4.3   EFFECT OF ELASTANE ON GARMENT FIT

Garment comfort is commonly associated with garment fit. Individual size, proportion, and shape are the most common issues among consumers in finding a correct fit garment. The difference between the leading brands and retailers' size range is another issue that bothers the consumer widely [7]. A perfect fit of the garment will aid good comfort to the wearer. A good fit of the garment can be reported based on the seam appearance and ensured by the absence of wrinkle, good position of the shoulder seam, and perfect vertical alignment of the seam to the floor. Likewise, a parallel hemline, relaxed neckline, smooth sleeve round at armhole represents a good fit [8]. A researcher reported a detailed analysis of the size difference among 42 different retailers and the sizes are not same [9]. This creates a serious issue, when the fit is considered. Further to add, due to the prevalent food habits the body mass index of individuals also increased or changed extremely and this also one of the serious in getting a good fit. Research reported that tight fit is a major issue and spoils the comfort of the garment. The study was performed among the women participants and they were asked to identify the fit of the garment at different body parts. The results showed that 79.4% of respondents reported an uncomfortable fit or tight fit at the bust region with the correct size garments [10]. The inconsistent body shape of the individual and the size difference in the products are due to the nonavailability of a standard size chart, which is the major reason for poor garment fit and comfort. Sometimes, this will lead the customer into frustration as they could not find a perfectly fitted garment at their favourite brand.

Garment fit becomes a trend due to the style feature of the garment. Hence, to obtain a good fit for different body sizes, it is necessary to use the garment made of stretch fabric, typically with elastane yarn can be used. Previous research work reported that the use of stretch wears will allow freedom of movement and provide enough comfort to the wearer. Though knitted fabrics are commonly used in the apparel industry, their application is not related to fit [11]. The most common method to introduce the stretch properties to the garment is through the use of elastane yarn in the fabric. However, these garments contain very less amount of elastane content in their structure, which makes the garment stretch and fit the contours of the body. But, it is also noted that due to the low elastane content, these garments will not create any compression effect or pressure on the body part as the compression garment does. Along with the stretchability, it is also necessary to mention the provision of freedom to the movement as a good sign of comfort [12]. In the analysis on the garments' comfort and aesthetic quality with elastane content, Geršak reported that the elastic properties potentially influence the garment's deformation kinetics and improve the fit. The inclusion of elastane content increases the elastic nature and has a direct relationship with the drape ability of the fabric. Their results reported a good correlation between the tensile elastic, bending, and shear elastic potential with the drape ability of the fabric and so the fit. Results of the study mentioned that the elastic potential of the fabric directly influences the drape ability of the fabric and so the fit characteristics. The elastic nature highly impacts the drape coefficient along with the crease depth due to their influence on the shear potential of the fabric [13].

Though no recent works were found in this area, few pieces of the literature showed the importance of elastane yarn in the garment fit. The structural and mechanical properties of the selected fabric are also one of the important parameters for garment fit. The fit was further influenced by the body measurement and garment construction properties [14]. Other researchers evaluated the handle and fit properties of the sari blouse made of cotton core-spun lycra yarns of different counts. They have related the fabric physical properties and low-stress mechanical properties with the garment fit through wear trial analysis. The perfection in the fit was evaluated using the fit attributes like ease measurement, the number of wrinkles, etc., subjectively using a 5-point scale. The results showed that a better fit was observed for the sample with a higher elastic stretch percentage. Out of the selected samples, fabric samples with higher stretch percentage (obviously more elastane percentage, the researcher did not report the content details), lower thickness, cover factor, and higher extensibility showed a better fit [15]. A particular sample showed a rank of "5" over all than the other samples due to their higher extensibility, lower shear, and bending rigidity value with superior total hand value. This ultimately increases the garment drape, and so the comfort and fit [15]. The researchers also analysed the effect of cotton and silk fabric as a warp with a cotton core-spun lycra as weft on fabric fit properties in terms of Indian tight-fitted garment saree blouse. They compared the results with 100% cotton fabric blouse material. The results showed that for both cotton and silk fabric, increment in the weft yarn count increases the elastane percentage, and so it showed a maximum stretchability. Hence, they reported that a woven fabric with finer yarn count had more stretchability. Out of the selected fibre type, silk fabric showed a higher stretchability compared to cotton fabric. A similar effect was also noted with the recovery percentage of the elastane fabric. Finer weft yarn increases the inter yarn friction and that restricts the fibre slippage and makes the yarn recover easily [16].

A similar study reported the fit analysis of silk/lycra blended yarn on garment fit. In the report, the effect of different structures on the garment fit with elastane included silk fabric. The researcher selected the Indian saree blouse as a target garment due to its higher fit requirements. Wearer's Perceived Fit Evaluation Analysis results showed that the highest garment fit was noted in the case of satin fabric due to its freedom of movement and easy-wearing properties. The researchers justified that the reason might be the higher extensibility of the particular fabric compared to the plain and crepe fabric. The second rating was given to crepe fabric due to its lower bending and shears rigidity values. The lowest rating was reported for the plain fabric. In the standardized fit rating scale analysis, it is noted that with lower folds sateen weave showed the highest fit rating followed by crepe and plain weaves. The main reason for the improved fit is due to the presence of elastane along with the silk yarn. They have confirmed that the improvement of the result in the fit through an ANOVA analysis. The mean fit rating for ease linear index, number of folds, and seam line deviation were above the rating of '4' in sateen and crepe weave blouses. By the results, they suggested the use of silk/elastane hybrid yarn application in saree blouses for improved fit [17].

Sertaç et al. investigated the effect of stretch on knit fabric made of 86% polyester and 14% elastane. The researcher analysed the influence of garment elastane stretch on the thermal behaviour of the apparel through the non-contact method. The virtual garment simulation results showed that the fabric's elastic nature affects the garment fit and also the thermal characteristics. They have reported that fabric extensibility was one of the important parameters which have more impact on the garment fit and thermoregulation of the body. Tighter fit will increase the temperature, and hence they suggested getting an optimum fit for improved thermal regulation [18]. The results of the study reported that an increase in the elastane content in the fabric converts the fabric as more stretchable which increases the garment fit. The available studies in the stretch garment fit evaluation mainly analysed the woven fabric over the knitted one.

4.4 FACTORS INFLUENCING THE FIT OF ELASTANE/STRETCH GARMENTS

Though the role of elastane influences the garment fit characteristics, the other fabric production parameters have their influence on the stretchable fabric fit properties. The factors influencing the fit of stretch fabric are provided in Figure 4.1.

4.4.1 NEGATIVE EASE

The fit of the garment was well defined in the pattern making and cutting process, as the measurements were decided over there. While the regular garments use positive ease allowance (Additional fabric with original body measurement), a tight-fitted or

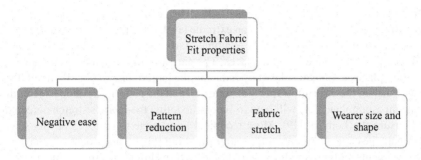

FIGURE 4.1 Factors influencing the stretch properties of elastane (stretch) fabric.

stretch fabric with elastane content required negative ease. Meaning that the measurements will be reduced based on the type of applications it proposed. As the pattern or fabric is prepared to a lesser measurement than that of the required body measurement, the amount of elastane present in the garment (stretch percentage) determines the fit of the garment. The difficulties arise during customization, the manufacturer may not predict and remove the excess fabric during this negative ease-making process. A high degree of mistakes and changes in fit level arises at this place and it significantly influences the wearer's comfort and fit. Garments with negative ease, are termed as proximal fitted garments by the researcher [19]. Based on its fitness level, Penelope Watkins reported different categories of the tightly fitted garments as form fit, cling fit, action fit, and power fit [19].

- The form-fit garments will have very few wrinkles and no stretch in them with minimal inherent stretch to ease the body movement.
- The second type is cling fit. It generally consists of stretch fabric that mainly conforms through the contour of the body. Though it does not impart any stress on the body, it clings to the body curves in a better way than the previously mentioned form fit garments.
- The third type is action fit garments. It specifically represents sportswear and exercise garments. These garments are available with different stretch levels. The action-fit garments are more stretch-oriented and its retracting stretch force helps in complete fit on the body components.
- Power fit garments are compression garments, which produce pressure on the body part and compress the flesh and change the shape of the wearer. These garments can be used as a whole garment or at specific locations to exert stress. These types of garments are used in professional sports and medical applications [19].

4.4.2 PATTERN DEVELOPMENT/REDUCTION

The second most important factor that influences the fit by negative ease is the pattern development process. In the case of professional sports applications, the garments will be custom made and they can be developed with the various panel to confirm the body contour of the sportsperson. However, when it comes to casual wear kind of tight fit garment, it addresses the mass market. Hence, these garments are expected to have all regular features along with an extreme level of fit. This creates complications to the pattern-making process. For example, darts are used in apparel to ensure fit and reduce the fabric buckling at places. The dart was also distributed completely to different places to enhance the fit by reducing the pattern proportionally in all directions. Further, the ease allowance was removed and the patterns developed according to the measurement. This will help in accommodating the fabric's stretch percentage and ensure the body fit [19]. More technical knowledge is required at this stageto eliminate the expected issues at the pattern-making stage. This process of converting measurements and making basic foundation patterns influences the garment fit in all garments. In the case of elastane stretch fabric, it has a significant role as it consumes negative ease in addition. To ensure a proper fit in the pattern making, a technician should have adequate knowledge of the stretch properties and amount of elastane in the particular fabric including its application. This will help them to understand the requirement and develop the pattern accordingly.

In a study, the researcher tried to standardize the woven stretch fabric through various extreme body movements. The study took 28 body measurements from five stretch positions and those measurements were used to develop the pattern. The researcher produced the reduced pattern by correlating the extension level of the fabric based on the following formula,

- Pattern reduction in the horizontal direction (cm) = Girth measurement (cm) × Fabric extension (%)
- Pattern increase in the vertical direction (cm) = Length measurement (cm) × Fabric contraction (%)

The study compared the fit of the garment that was produced through this modified method and conventional method. The subjective rating results revealed that the proposed method fits better than the conventional pattern-making method. However, the authors suggested the need of standardizing the pattern drafting procedure [20]. Other researchers developed a calculation method to draft the pattern for the stretch garment. From previous literature, the authors developed a formula for the pattern measurement reduction in both width and lengthwise using a gridwork that the researcher developed. The details of the grid work on the pattern can be accessed in the literature [21]. The formulae were obtained by different practical tests performed at the various garments with a different stretch level in the industry.

4.4.2.1 The length reduction formula

a. Total pattern length reduction = Distance from neck to hip × Higher or lower percentage of fabric stretch by a load test method.

b. (Total pattern length reduction/2/)3) = horizontal reduction units used in each of three locations on the pattern

In this method, the researcher calculated the total length reduction and divided it by two to get 50% of the measurement. The previous research reported a skin stretch of approximately 50% in the areas of knee, elbow, and seat, which are reported as maximum stretch [21]. Hence, to incorporate the adequate ease researcher also divide it further by three to establish a correct amount of reduction through the body.

4.4.2.2 The width reduction formula

a. Half body width reduction = Distance between the centre front to centre back at bust level X Higher or lower percentage of fabric stretch by a load test method.
b. [Half body width reduction/2]/2 = Total reduction in each 1/4th of body
c. [Total reduction in each 1/4th of body/4] = Neck/hip reduction unit
d. Neck/hip reduction unit × 3 = Shoulder/hip reduction unit

As mentioned in the vertical pattern, detail in this formula initially identifies the half body reduction and multiplied with stretch percentage. The quarter body measurements were identified and fitted at different grid levels [21] of the garment to avoid distortion of the garment. In this way, researchers developed a pattern and stitched leotards. Through visual examination, a set of standard checkpoints were verified and the method was found suitable for developing stretch fabric patterns [22].

When the stretch fabric patterns are concerned, the textbooks exist in the pattern drafting area that refers to patterns drafted smaller than the actual size. The textbook written by Aldrich refers to two different types of patterns for stretch fabric. The first one is represented as an "easy fit block" and the second one is called a "close-fitting body block". The first one, the easy fit block, was developed based on the burst measurement and the second based on body dimensions [23]. The other textbook by Armstrong reported a body block reduction at waist and hip reduction by keeping the bust measurement without altering it, for stretch fabric [24]. Dove reported a method for stretch fabric by reducing woven basic block. In this method, the author used back body block as a template and reduced the ease allowance, and then applied the body measurement reductions [25]. While discussing the stretch fabric pattern reduction, Aldrich and Aldrich reported the use of subjective stretch measurement. They detailed the woven stretch fabric pattern reduction steps. The reduction needs to be performed after measuring the stretch level after a hand stretch or by a simple load extension stretch method. Based on the extension percentage, the authors suggested the following points for pattern reduction to achieve a good body fit:

• A reduction of 5% in all widthwise measurements for every 10% of subjective stretch measurement.
• An increase of 2% for every 5% increment in the stretch under load, for vertical measurements [26].

The main drawback of this method is that these modifications are highly influenced by the stretch measurement by subjective analysis. Disher reported a set of procedures to prepare a stretch pattern for the elastane fabrics. He reported that for a fabric with a higher stretch percentage, the patterns must be reduced down to the sizes and should be adopted for a short wear trial to ensure fitness. The reduction in the size of the pattern should be based on the level of the stretch associated with the stretch fabric. After the wear trial, the researcher also reported to wash or dry clean the fabric and check for the measurement changes. This will give a wide range of modifications in the measurement. He suggested taking this step before adopting the pattern or during the bulk cutting in the industry [27]. Though each fabric must be evaluated through the mentioned steps, the mistakes cannot be eliminated due to the subjectiveness of the steps involved in this process.

Another expert in the field listed a similar and simple method for the pattern reduction of stretch fabrics. These formulas were developed based on their personal experience in the particular field [28]. For the bust, waist, and hip areas, the pattern can be reduced using this formula:

- Total amount of comfortable stretch = {Body Measurement × 0.3 (i.e. 30%)} + Body Measurement.

For the front rise region of the garment, the percentage is reduced to a level of 10% by predicting the lower vertical stretch of the fabric. The formula is

- Total amount of comfortable stretch = {Body Measurement × 0.2 (i.e. 20%)} + Body Measurement [28]

4.5 FABRIC STRETCH PERCENTAGE

Understanding the fabric stretch percentage is very crucial to maintain the perfect fit of the stretch fabric. In industry, the fabric stretch level is measured in three different subjective ways [19].

1. The most common method used is measuring the stretch level with a scale. This method is completely subjective and the technicians' knowledge plays a vital role in defining the stretch percentage.
2. The second method represents the modelling of the stretch fabric directly on the body form of the target size. This is also one of the ways of subjective analysis; however, the error percentage will be less than the previous one.
3. The last method is the most generic way of developing stretch patterns, that most of the industries use still. In this process, the industry uses one down-size pattern (lower size) with the general assumption the stretch level of the fabric will automatically adjust the need for body parts [19].

These methods are complicating the stretch garment fit estimation. The literature survey also did not show any textbooks with a solid way of developing patterns for stretch fabrics. In one such material, the author estimated the requirements of fit through a mathematical model and reported that reducing the circumferential measurement to 20% and vertical measurement to the level of 20%–25% will yield a better fit. However, it was also reported by the authors that this formula will not fit all the cases and subjective alterations required with respect to fabric stretch and subject [29]. Similarly, other researchers also reported the complications associated with the ease and pattern drafting of stretch fabric [30]. Though many studies addressed the subjectiveness of the stretch percentage measurement, very few researchers addressed the objective measurement of fabric stretch percentage. A research study used a test method that was similar to ASTM D-2594, 1982, to evaluate the stretch percentage of the fabric. In this method researcher, developed a 40cm length and 20 cm wide fabric. Each side was equally hemmed and one side hanged in a hanger. The working area of the sample was maintained as 20 cm wide and 20 cm lengths. In this method, at the bottom side, opposite to the hanger, a 500 g load was applied and allowed to stay up to 30 minutes. The extension level of the fabric was measured using the formula.

Percentage of fabric stretch = 100 {(Distance between the 20 cm marks before extension − Distance between the 20 cm mark line under load)/ Distance between the 20 cm mark line under load.

In this way, the stretch of the fabric calculated in the course and wales wise and the combined total stretch percentage were calculated as an average of these percentages [22]. In their results, they compared different fabrics with different elastane percentages and reported that though the overall stretch percentage of two different fabrics noted less, their coursewise and waleswise stretch noted significantly different with respect to the fabric structure and elastane percentage. Hence, to develop perfectly fitted garments, it was necessary to understand and standardize the both course, wales, and overall stretch of the garment [22]. Another researcher also mentioned that the commonly adopted method is not satisfactory due to its higher subjective nature. In this method, the fabric was stretched against a ruler and the sample was categorized as low-, medium-, and high-stretch fabrics. Theoretically, the stretch percentage is the point at which the stretch fabric has reached its maximum extension without deforming hard yarns and fibres [31]. This stretch test estimates two parameters subjectively, (i) the amount of stretch force applied on the fabric and (ii) the point at which the fabric visually stretched. These two aspects are totally judgemental based on the technical knowledge and expertise of the technician who performs the analysis. The researcher also reported that the mechanical test methods adopted to examine the degree of stretch are often influenced by the sample length, width, and the amount of load applied [32]. The common standards analyse the fatigue of the specimen, determining the extension at a specific force, modulus, residual extension, fatigue set, and elastomeric thread break which are not suitable for the stretch pattern making. In line with the previous researcher, the researcher also reported that the averaged extension percentage figures reported by manufacturers (Average of course direction and wales direction extension percentage) often lead to the problem and are inadequate to design the stretch pattern. [32]

4.6   BODY SHAPE AND SIZE STANDARDIZATION

Human body shape and size are other important factors that have a significant impact on the fit of the apparel. It is very difficult to determine the apparel-to-body-fit relationship with some generic measurements. More details on the measurements will provide adequate fit. Though the existing brands use sizing systems to designate apparel of different sizes and provide a size label, it does not define anything on fit. This creates difficulty in adapting a particular size without performing a wear trial. Due to the difference in the body shape, proportion, and postures within the same size or same circumferential measurement category, it is very difficult to standardize the size [30]. This is one of the major reasons for fit issues in garments with non-stretch or stretch characteristics. We can practically see many of us not fitting into the size range provided by the major brands. A large number of customers find themselves between size range due to their body shape. Another important issue related to the size range offered by the retailer is, they are not interested in the accurate measurements of their customers due to the complication and cost associated with it. Retailers wanted to address the majority of the consumers; hence, they always prepare sizes that can accommodate large people with small variations. Sometimes, we can feel this while one brand's particular size can provide a perfect fit, the same size in another brand provides a tighter or looser fit. These are all the major reasons why the sizes and body shapes are not standardized still.

The reason behind this size classification is, the brands like to address a wide range of consumers through alpha or letter sizing (S, M, L, etc.), instead of numerical sizing. This is mainly because letter sizing gaps are almost double the numerical sizing. Further to complicate, the size difference is huge among the different countries like the US, Europe, Asia, etc. The main issue here is the lack of availability of anatomical data in the world. Until recent times, no study recorded the data of human body size and shape. In 2005, the body form manufacturer Alvanon conducted a study across the US along with NC state university. The findings of the results reported that only 8% of the study population had the hourglass-shaped body, which was assumed to be the standard body shape (by the government and the majority of the retailers) from the 1960s [33].

For example, researchers reported that the body proportion changes are also one of the important aspects of stretch garment fit. Concerning vertical measurements, it is possible to have the same amount of bust, hip, and waist measurements, though they are from different height categories like small, medium, and tall humans. Researchers reported these changes will reflect in the torso proportion of these types of people. Further, with the width measurements considered, even in the same size of a torso, different widths in the front and backside are also possible. Possible to have a different measurement in the side to side in the front (wide) than the backside (short). The researcher reported that the existing system of measurement and pattern drafting should be reformed in the case of stretch fabric applications [34].

The details presented in this section summarize the necessity of the future works that should be performed in this area. As discussed in the section, it can be clearly seen that a lot of fabric properties were analysed by a different researcher with different elastane percentages, elastane insertion percentages, and also with both knit and woven fabrics. However, these properties are mainly associated with fabric properties, and the fundamental-fit properties come with the pattern making and the type of measurements used in the sewing process. In this aspect, the factors that influence the stretch garment fit were also discussed in this section. It can be seen that there is a wide scope in this area for future research to standardize the pattern reduction and stretch percentage estimation. Though new studies were performed in the recent time to standardize the body shape and size, a lacking in the previously mentioned two issues remains still as fresh in the previous decade.

4.7 ELASTANE IN GARMENT COMFORT

Clothing comfort is one of the complicated aspects of clothing. It does not have a direct measure to quantify. However, it is always one of the essential requirements for clothing. During the wear phase, irrespective of the type of apparel or environmental condition, a human being always expects comfort from the clothing. Due to nature and variability, clothing comfort is not defined by a single parameter. The comfort aspects of the clothing are generally based on the permeability characteristics like air, water vapour, heat, and moisture transmission. Likewise, the moisture handling abilities of the textiles namely, wicking, absorbency, liquid moisture transmission, and drying characteristics all significantly contribute to the clothing comfort of the textiles. This section of the chapter summarizes the influence of elastane yarn usage in fabrics on the comfort characteristics of textiles. The fundamental requirement of the elastane is to increase the stretchability of the textile material and so the reliance on the textile can be increased. This is also noted as one of the important aspects of the garment or apparel fit and comfort. However, the usage of elastane yarn inside the textile structure also has negative effects on the comfort characteristics of the textile material. Figure 4.2 represents the factors influencing the comfort characteristics of clothing.

Knitted fabrics are generally used in tight-fit garments due to their higher stretch, recovery rate, and comfort. However, it is necessary to include the elastane yarns in the knit structure to increase the comfort requirements of knitted fabric at various

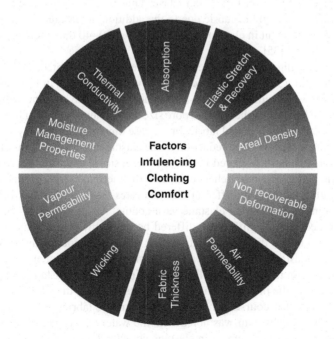

FIGURE 4.2 Factors influencing the clothing comfort.

physical activities. The use of elastane in the fabric structure increases the application potential of the fabric in multifold due to its higher stretch and recovery rate with required comfort. The use of elastane in the fabric structure, specifically in the case of knit structures, alters the physical properties to a greater extent. The use of elastane in a knitted fabric affects the GSM, air permeability, pilling, spirality, loop length, course spacing, wale spacing, course/cm, wales/spacing, stitch density, and bursting strength and shrinkage [6]. All these properties are either alone or by combination influences the comfort characteristics of the knitted fabric.

Selin Hanife Eryuruk and Fatma Kalaoglu evaluated the properties of knitted fabric with two different types of elastane plating techniques namely half plating and full plating. The results of the study showed that the increment in the elastane percentage, from half to full plating, increased the elastic properties of the knitted fabric. Since the elastane was inserted in the weft wise, the extension of the knitted fabric was noted higher in the case of widthwise direction. The most important property that influences the fabric's comfort is the recovery property of the fabric. Research results reported that the linear density of the elastic yarn had a significant impact on the recovery properties of the elastane knitted fabric. On the increment in the elastane yarn linear density from 20 to 40 dtex in both the half- and full-plating fabrics, the full-plated knitted fabric showed a higher elastic recovery value than the half-plated fabric. This represents more elastic percentage in the fabric increases the elastic recovery of the fabric [35]. On the evaluation of drape coefficient values between full-plated and half-plated elastane knitted fabric, the results showed a higher drape coefficient for full-plated fabric. This might be attributed to the higher tight structure and weight of fabric due to the addition of elastane. Similarly, it also noted that the increment in elastane yarn count from 20–40 dtex increased the drape ability of the fabric [35].

While comparing the air permeability values, the half-plated fabric showed a higher value than a full-plated fabric. Similarly, the increment in the elastane yarn denier also showed a reduction in the air permeability of the elastane knitted fabric. This was noted because of the reduced pore size and tighter structure of the full-plated fabric compared to a half-plated fabric. In the case dimensional stability values, there was a reduction noted with the reduction in the elastane yarn count. Similarly, on the evaluation of half and full-plated fabric, the dimensional stability is better for half-plated fabric [35]. Other researchers evaluated the thermal comfort properties of the elastane knitted fabric with different elastane rates (without elastane, half plating, and full plating) and different elastane yarn counts. Air permeability values of the elastane knitted fabric significantly differed between different elastane counts and feed rates. Out of the different counts tested, a higher air permeability was noted with 40 denier elastane yarn than 20 and 60 deniers. In the case of elastane feed rate, there is a reduction in air permeability with an increment in elastane feed rate. This was mainly due to the increment in the fabric density due to the addition of elastane. Out of all the samples, courser denier, full-plated knitted fabric showed lower air permeability. Similar behaviour was noted with the water vapour permeability of elastane knitted fabric. An increment in elastane count and feed rate reduced the fabric porosity of the fabric and so reduces the water vapour permeability. Concerning the thermal conductivity values, the elastane yarn with 60 deniers showed a difference in thermal conductivity. Whereas the lower count elastane yarn did not show any significant difference. When compared to the fabric without elastane and half-plated fabric, the full-plated fabric showed a higher thermal conductivity value. This might be attributed to the higher thickness of the structure [36].

The use of elastane yarn in knitted fabric restricts the air permeability of the fabric compared to natural fibre knitted fabrics. Further to add, the use of elastane reduces the wettability of the knitted fabric. Hence, during the wear, the elastane knit structures poorly absorb sweat from the skin and reduce the thermo-physiological wear comfort. Based on this, the effect of the use of core-spun elastane-cotton blend yarn on thermo-physiological wear comfort was analysed by Manshahia and Das. The results were analysed for the properties such as air permeability, water vapour permeability, thermal resistance, wicking, and absorption. They have reported that the increment in elastane percentage and elastane content made the fabric more compact and thick, which increased the thermal resistance and reduced the permeability values due to the reduction in pore size. Similarly, increment in elastane stretch and elastane percentage reduced wicking and absorption properties of elastane knitted fabric. The effect of the yarn twist had a positive effect on the moisture and heat transfer through the fabric [37].

Other researchers analysed the effect of elastane proportion on the comfort properties of the denim fabric. For the research, they have developed denim casual fabric with different elastane ratio percentages only in the weft direction. The results showed that an increment in the elastane percentage reduced the tensile strength and tearing strength of the denim fabric. Similarly, the increment in elastane percentage made the denim fabric stiffer than the non-elastane fabric. On the evaluation of fabric bagging, the researcher found that the increased elastane percentage reduced the permanent elastic deformation due to higher elasticity. This is an advantageous effect due decrease in the bagging deformation. But at the same time, the researcher cautioned that though a higher elastane percentage favoured positive properties for comfort, a higher percentage of addition beyond a certain level increases the stiffness of the fabric [38]. Likewise, the thermo-physiological wear comfort of the synthetic sportswear fabric with elastane content was also analysed by researchers. In the case of sportswear, due to the wearer's excessive sweating, the sweat transmission from skin contact to the other side is an important aspect for comfort. Likewise, the transmission of heat through the structure is also one of the main parameters. Hence, the research analysed the air permeability, OMMC, permeability index, and thermal characteristics. The results of the research indicated that the increment in elastane linear density reduced the air permeability values due to its more compact structure. Concerning the thermal transmission behaviour, the researchers reported a poor heat resistance with coarser elastane linear density than a fine elastane. Due to the hydrophobic nature of the elastane filament, coarser elastane fibre showed a reduction in moisture vapour transmission and permeability index. These changes are attributed to the increased fabric tightness and reduction in inter yarn space than the fine elastane knitted fabric. The researcher also noted a similar effect in the case of water absorption and wicking properties with different elastane linear densities.

In a study, cotton and cotton/elastane blended yarns were compared for their physical properties. The addition of elastane increased the weight of the sample as the loop length decreases. As reported earlier, the findings showed lower air permeability of cotton/elastane fabric than 100% cotton fabric. The changes can be correlated with the loop length and elastane percentage in the fabric. The findings revealed that the inclusion of elastane did not show any impact on pilling property. Further elastane fabric showed an increment in the stitch density and bursting strength due to the elastane content and there was a reduction in shrinkage. All these parameters change the dimensional properties and so comfort characteristics [39].

The thermal comfort properties of denim wear were evaluated by Selin Hanife Eryuruk [40]. The researcher used 100% cotton fabric and 98% cotton and 2% elastane denim structures for thermal comfort. The results of the study showed that physical properties like tear resistance, stiffness, and drape values were affected by the addition of elastane in the denim structure. The inclusion of elastane reduced the tear resistance of the denim fabric but at the same time, it reduced the shear stiffness and bending rigidity values of the fabric compared to 100% cotton denim fabric. Further, the addition of elastane showed a significant increment in the fabric resilience value and smooth surface than cotton denim. Concerning the thermal properties, the results showed a higher thermal resistance value with denim fabric with elastane fabric than 100% cotton fabric. Hence, the thermal conductivity, thermal absorptivity, and thermal diffusivity values were noted less in the case of elastane fabric as compared to nonelastic fabric. The researcher concluded that the addition of elastane fibre increased the handle and comfort properties of the textile fabric [41]. Dunja Sajn Gorjanc et al. analysed the influence of the addition of elastane content on the water vapour permeability and thermal conductivity of the cotton fabric. It revealed that the inclusion of elastane yarn in the weft direction increased the warp density of the fabric. This increment in the yarn density reduces the water vapour permeability of the elastane-incorporated fabric up to 20% than the normal cotton fabric. Similarly, in the case of thermal conductivity, the addition of elastane resulted in a similar effect. The increment in the elastane content significantly increased the thermal resistance of the woven fabric [42].

Bilal Qadir et al. analysed the effect of elastane denier and draft ratio on the mechanical and recovery properties of the woven fabric. In this study, researchers used elastane in the weft direction alone. The findings showed that the addition of elastane in the weave structure increased the areal density of the woven fabric. An increment in areal density was noted with an increase in elastane denier and draft ratio. This is mainly due to the weftwise contraction of the fabric due to elastane inclusion. This was noted higher with a higher elastane percentage and draft ratio. As a consequence, the increment in elastane percentage and draft ratio also increases the fabric tensile and tearing strength. In the case of comfort properties, the fabric's ability to stretch and recovery were observed. The results indicated that the stretchability of the fabric increases with the elastane denier percentage. An increment in the denier percentage increases the elastane content on the fabric and so it increases the stretchability. Likewise, the higher draft ratio also increases the stretch property. The higher draft ratio reduces the elastane content due to the higher retraction force of the elastane yarn. In the case of recovery property, higher denier elastane showed a higher elastic recovery than the lower denier yarns. However, a negative effect was noted with the increment in the draft ratio percentage. A higher draft ratio reduces the fabric stretch and recovery due to the reduction in the elastane content [43]. Other researchers measured the effect of elastane content on the non-recoverable deformation of the woven fabric. The non-recoverable deformation of the fabric represents the difference between the total deformation and the recoverable deformation of the fabric under stress-strain curve. The researcher evaluated the effect of different fabric structures and elastane proportion in the weft direction on non-recoverable deformation. The results of the study showed that the addition of elastane increased the breaking extension and reduced the elastic modulus significantly. As a result, the inclusion of elastane content reduced the non-recoverable deformation up to 20% compared to the cotton plain-woven fabric. In the case of twill weave, the inclusion of elastane reduced the non-recoverable deformation from 20% to 5% for different elastane content [44].

In summary, the inclusion of elastane in the fabric increases the areal density of the textile material. This subsequently increases the weight and thickness of the fabric upon the addition of elastane yarn. As the stretch properties increase with elastane percentage, the elastane yarn tightens the structure and so reduces the intra yarn spaces or porosity of the fabric. This results in air and moisture permeability of the

TABLE 4.1
Effect of elastane content on the comfort-related fabric characteristics

S. No.	Fabric characteristics	Influence of elastane addition	Effect on comfort
1	Thickness	Increases as the structure compactness increases	Increases stiffness and reduces the comfort
2	Areal density	Increases as the structural tightness increases	Reduces the comfort as the thickness and compactness increases
3	Air permeability	An increase in thickness and areal density reduces the pore size and spaces between successive loops	Reduces the comfort
4	Wicking	Reduces due to the hydrophobic nature and compact structure	Reduces the comfort
5	Absorption	Reduces due to the hydrophobic nature	Reduces the comfort
6	Vapour permeability	Reduces the water vapour permeability	Reduces the comfort
7	Thermal conductivity	Increases the fabric density and weight and reduces the thermal conductivity	Reduces the comfort
8	Moisture-management properties	Reduces due to the increase in areal density and poor absorbency	Reduces the comfort
9	Non-recoverable deformation	Reduces the deformation	Increases the comfort
10	Elastic stretch and recovery	Increases the stretchability and recovery	Increases the comfort

elastane-incorporated textiles. When the geometrical properties are considered, in both the knit and woven fabric, inclusion of elastane yarn increases the warp and weft or the yarn density in the fabric. In both cases, the width of the fabric decreases with increased elastane content. On the positive side, the elastane content increases the stretchability and reduces the non-recoverable deformation and bagging effect on the fabric. A higher percentage of elastane inclusion significantly causes many negative effects to the wearer clothing comfort like poor air permeability tighter and stiffer fabric. A smaller percentage of plating or pick density increases the elastic nature and so reduces the elastic modulus of the fabric. This will reduce the non-recoverable deformation and higher surface smoothness. And so it increases the wear comfort up to a certain level for the knitted and woven fabrics. Table 4.1 consolidates the effect of elastane inclusion on the different comfort parameters and summarizes it.

4.8 SUMMARY

This chapter evaluates the role of elastane in garment fit and comfort characteristics. The first section details the essential properties of stretch fabric produced from elastane. Further, the chapter discussed the factors influencing the garment fit. The results showed that the lack of standard methods to determine the fabric stretch and pattern reduction based on the stretch level was the ideal issue in influencing the stretch garment fit. The knowledge about the body shape, proportion, and size was also another factor that had the least attention while discussing the garment fit. When the garment comfort aspects are concerned, very few properties like stretch and recovery properties alone provide a better comfort characteristic when the elastane yarns are included in both knit and woven fabrics. Apart from that, all other properties like air and moisture permeability, thermal conductivity, and moisture handling properties including liquid absorption, spreading, and wicking properties reduce significantly. The reduction in these properties is a sign of warning at extreme applications like sports, at an excess sweating, a higher level of elastane percentage may put the wearer at discomfort. Hence, it is advisable to select the level of elastane content according to the application requirements.

REFERENCES

1. Slater, K. (1986). The assessment of comfort. *Journal of the Textile Institute,* **77**(3), 157–171.
2. Stamper, S., Sharp, H., & Donnel, L. (1991). *Evaluating apparel quality.* New York: Fairchild.
3. Brown, P., & Rice, P. (1998). *Ready-to-wear apparel analysis* (2nd ed.). NewJersey: Merrill-Prentice Hall.
4. Alexander, M., Connell, L.J., & Presley, A.B. (2005). Clothing fit preferences of young female adult consumers. *International Journal of Clothing Science and Technology,* 17(1), 52–63.
5. Rasband, J.A. (2002). *Wardrobe strategies* (2nd ed.). New York: Fairchild.
6. Minah Thembi Nkambule. (2010). Apparel sizing and fit preferences and problems of plus-size swazi working women. Master's thesis, Department of Consumer Science, University of Pretoria.

7. Otieno, R.B. (2011). *Improving apparel sizing and fit. Advances in apparel production.* Cambridge: Woodhead Publishing Limited.
8. Betzina, S. (2003). *Fitting basics. Fast fit: Easy pattern alteration for different figures.* Newtown, CT: Taunton Press.
9. Tanya Dove. (2016). Stretch to fit – made to fit. *International Journal of Fashion Design, Technology and Education,* 9(2), 115–129. DOI: 10.1080/17543266.2016.1167252
10. PisutLenda, G., & Connell, J. (2007). Fit preferences of female consumers in the USA. *Journal of Fashion Marketing and Management,* 11, 366–377.
11. Laing, R.M., &Sleivert, G.G. (2002). Clothing, textiles, andhuman performance. *Textile Progress,* 23, 1–104.
12. Ashdown, S.P. (2011). Improving body movement comfort. In Apparel. In G. Song (Ed.), *Improving comfort in clothing* (pp. 278–301). Cambridge: Woodhead Publishing Limited.
13. Geršak, J. (2004). Study of relationship between fabric elastic potential and garment appearance quality. *International Journal of Clothing Science and Technology,* 16(1/2), 238–251. http://dx.doi.org/10.1108/09556220410520513
14. Rodel, H., Schenk, A., Herzberg, C., & Krzywinski, S. (2001). Links between design, pattern development and fabric behaviors for clothes and technical textiles. *International Journal of Clothing Science and Technology,* 13(3/4), 217–227.
15. Nirmala Varghese, & Thilagavathi, G. (2013). Development of woven stretch fabric for comfortably fitting sari blouses and analysis of fit. *International Journal of Fashion Design, Technology and Education,* 6(1), 53–62. DOI: 10.1080/17543266.2013.764022
16. Nirmala Varghese, & Thilagavathi, G. (2015). Development of woven stretch fabrics and analysis on handle, stretch, and pressure comfort. *The Journal of the Textile Institute.* 106(3), 242–252. DOI: 10.1080/00405000.2014.914652
17. Nirmala Varghese, & Thilagavathi, G. (2016). Handle, fit and pressure comfort of silk/hybrid yarn woven stretch fabric. *Fibers and Polymers,* 17(3), 484–494.
18. Sertaç Güney, Hilal Balci, & İbrahim Üçgül. (2019). Investigation on the effect of garment fit on thermal transfer performance of clothing by combining non-contact measuring tools. *International Periodical of Recent Technologies in Applied Engineering,* 1, 9–12. DOI: 10.35333/porta.2019.18
19. Penelope Watkins. (December 2011). Designing with stretch fabrics. *Indian Journal of Fibre and Textile Research,* 36, 366–379.
20. I-Chin, D. Tsai, Carol Cassidy, Tom Cassidy, & Jinsong Shen. (2002). The influence of woven stretch fabric properties on garment design and pattern construction. *Transactions of the Institute of Measurement and Control,* 24(1), 3–14.
21. DuPont International. (1975 or 1978). *A practical guide to stretch.* Geneva, Switzerland: DuPont.
22. Ziegert, B., & Keil, G. (1988). Stretch fabric interaction with action wearables: Defining a body contouring pattern system. *Clothing and Textiles Research Journal,* 6(4), 54–64. doi:10.1177/0887302X8800600408
23. Aldrich, W. (2008). *Metric pattern cutting for women's wear* (5th ed.). Oxford: Blackwell Publishing.
24. Armstrong, H.J. (2013, July 23). *Patternmaking for fashion design* (5th ed.). New Jersey: Pearson.
25. Dove, T.L. (2013). *A technical foundation, women's wear pattern cutting.* London: Austin Macauley.
26. Aldrich, W., & Aldrich, l. (1996). *Fabric fonn and flat cutting* (pp. 138–141). Oxford: Blackwell Science.
27. Disher, M. (1980). All about stretch, *Manufacturer clothing,* DuPont, London. (pp. 47–53).
28. Brittani, B. (2020). How to choose a size with negative ease. https://untitledthoughts.com/blogs/intro-to-knits/how-to-choose-a-size-with-negative-ease

29. Pratt, J., & West, G. (2008). *Pressure garments a manual on their design and fabrication* (pp. 22–24; 32–33). Oxford, UK: Butterworth Heinemann Ltd.
30. Shoben, M. (2008). *The essential guide to stretch pattern cutting: dresses, leotards, swimwear, tops and more.* London: ShobenFashion Media.
31. Murden, F.H. (1966). Elastomeric thread Review (II): Elastomer and fabric test method. *Textile Institute and Industry*, 4, 355–358.
32. Penelope, A.W. Custom fit: Is it fit for the customers. Working paper, London College of Fashion, University of the Arts, London. https://ualresearchonline.arts.ac.uk/id/eprint/1032/1/Custom%20Fit.pdf
33. Gribbin, E.A. (2014). Body shape and its influence on apparel size and consumer choices. In Book. In Marie-Eve Faust & Serge Carrier (Eds.), *Designing apparel for consumers, the impact of body shape and size.* Cambridge, UK: Woodhead Publishing Limited in association with The Textile Institute Woodhead Publishing Limited.
34. Watkins, P.A. (1999). Design for movement: Block pattern design for stretch performancewear. Doctoral thesis from De Montfort University, UK. https://dora.dmu.ac.uk/handle/2086/19315
35. Selin HanifeEryuruk, & Fatma Kalaoglu. (2016). Analysis of the performance properties of knittedfabrics containing elastane. *International Journal of Clothing Science and Technology*, 28(4), 463–479. https://doi.org/10.1108/IJCST-10-2020
36. Gözde Ertekin, Nida Oğlakcioğlu, & Arzu Marmarali. (2018). Strength and comfort characteristics of cotton/elastane knitted fabrics. *Tekstilve Mühendis*, 25(110, 146–153. https://doi.org/10.7216/1300759920182511010
37. Manshahia, M., & Das, A. (2014). Thermo-physiological comfort of compression athletic wear. *Indian Journal of Fibre and Textile Research*, 39, 139–146.
38. Nilgün Özdil. (2008). Stretch and bagging properties of denim fabrics containing different rates of elastane. *Fibres and Textiles in Eastern Europe*, 16(1), 63–67.
39. Manshahia, M., & Das, A. (2014). Thermophysiological comfort characteristics of plated knitted fabrics. *The Journal of the Textile Institute*, 105(5), 509–519. https://doi.org/10.1080/00405000.2013.826419
40. Abu Yousuf Mohammad Anwarul Azim, Kazi Sowrov, Mashud Ahmed, Rakib Ul Hasan, H.M., & Md. Abdullah Al Faruque. (2014). Effect of elastane on single jersey knit fabric properties – physical & dimensional properties. *International Journal of Textile Science*, 3(1), 12–16. https://doi.org/10.5923/j.textile.20140301.03
41. Selin HanifeEryuruk. (2019). The effects of elastane and finishing processes on the performanceproperties of denim fabrics. *International Journal of Clothing Science and Technology*. 31(2), pp. 243–258. https://doi.org/10.1108/IJCST-01-2018-0009
42. DunjaSajnGorjanc, KrsteDimitrovski, & MatejaBizjak. (2012). Thermal and water vapor resistance of the elastic and conventional cotton fabrics. *Textile Research Journal*, 82(14), 1498–1506.
43. Bilal Qadir, Tanveer Hussain, & Mumtaz Malik. (2014). Effect of elastane denier and draft ratio of core-spun cotton weft yarns on the mechanical properties of woven fabrics. *Journal of Engineered Fibers and Fabrics*, 9(1), 23–31.
44. Dunja Sajn Gorjanc, & Matejka Bizjak. (2014). The influence of constructional parameters on deformability of elastic cotton fabrics. *Journal of Engineered Fibers and Fabrics*, 9(1), 38–46.

5 Elastane in medical applications

5.1 INTRODUCTION

The fabrics that are plated or containing some proportion of elastane yarns show higher elasticity, dimensional stability, and a higher degree of recovery than a fabric without elastane content. The main application of elastane in the medical field is a compression garment. The purpose of the compression garment is to create some pressure on a particular portion of the body for medical reasons. Applications of a compression garment on the human body have some positive effects like controlling blood pressure, reducing muscle strains, and sprains. In some cases, compression garments are also used to accelerate the healing process of different wounds like burn wounds, scars, and issues like deep vein thrombosis. When compared to the sports application, the compression garments used in medical applications are more specific to the requirement of the patients. They are manufactured based on the requirements of the individual patient's needs and also based on the application area like the sleeve, stockings, gloves, etc. The medical compression garments are worn on the "instructed" time and whenever there is a requirement of such treatment to the patients.

Compression garments are generally developed in a lesser circumference size than the body part where its application is aimed. The garments are particularly designed in this manner to achieve a compression effect by stretching yarns in the fabric. Once a person wears a compression garment, it stretches due to the yarn in the fabric is in compressed form and immediately, this stretch creates a force on the applied body part as the stretched yarns try to return to their original position. It can be understood that the pressure inserted by a compression garment is mainly associated with the elastane yarn in it. The amount of pressure created in the body part is basically due to the resilience effect of the elastane yarn. In general, medical compression garments are produced using ground yarn and inlay yarn. The ground yarn provides the thickness and stiffness to the compression garment, whereas the inlay yarn provides stretchability based on the elastic yarn used. Most of the time pure elastic yarns like lycra or spandex will be used to produce compression garments. Sometimes, core or cover spun yarns are also used as inlay yarn. In knit structure, the elastic yarn can be used in various forms namely as an inlay or as a float, or as a plated in the structure [1]. The insertion of the inlay yarn can be performed based on the requirement, either in every course of yarn or after certain courses. This elastic insertion decides the compression ability by altering the compactness and the strength of the compression fabric. Generally, the compression level of the garment is based on the type of elastane yarn used and its properties like elastic modulus, stretchability, and type of cover yarn (if used). Elastic knitted fabric produced from the cut and sew method

DOI: 10.1201/9780429094804-5

often have several disadvantages like the presence of seam and stretch issues, pressure variation at seams, etc. Hence, several researchers proposed shaped knitting on flat or circular knitting machine-oriented products [2]. Perfect fitting due to the anatomical shape development, higher compression effect, and integration of pads and support effects to improve the blood circulations are potentially possible with flat or circular knitted products [2].

The application of compression or pressure on a particular part of the body is generally known as compression therapy. The pressure can be applied over a selected point and ensures that the pressure on the inflamed site is less and immobile during the activity. Similarly, active compression therapy is used to apply pressure and enhance the blood flow in the case of edema and varicose vein issues. An average pressure of 25 mmHg is more than enough to speed up the blood flow in the veins. The interface pressure induced by compression or support garment is defined by Laplace's equation as [3],

$$P = \frac{TnK}{CW}$$

where,

- P is the pressure exerted by the bandage (mmHg);
- T is the tension applied by the bandage (kgf);
- n is the number of layers;
- C is the limb circumference (cm);
- W is the width of the bandage (cm); and
- $K = 4630$.

Table 5.1 represents the different possibilities of typical knitting constructions utilized in the production of elastic fabrics which are commonly used in medical applications [4].

As the compression ability of the fabric mainly depends on the mechanical properties of the fabric, it is essential to monitor the linear density, draw ratio, type of covering, covering rate, and twist per metre of knitting yarns. These parameters largely alter the overall mechanical properties of the compression garments. The higher linear density of the inlay yarn makes the compressive fabric thicker and more stable for long-term use. Further, the yarn or material used depends on the pressure requirements of the end-user. Figure 5.1 represents the construction methods of the elastic fabric and the dark line reports the elastic yarn passage in the structure.

Stiffness and elasticity are the two common mechanical properties that have a direct influence on the pressure generated in a compression garment. The stiffness of the textile material represents the bending resistance of the material. It was noted that the stiffness of the material has a significant impact on the compression properties of the compression garment. The stiffness of the garment was reported as the ability of the textile material to reflect the muscle expansion during contraction. In compression therapy, the Static Stiffness Index (SSI) and the Dynamic Stiffness Index (DSI) are used to assess the stiffness of the compression textiles [5]. The first parameter

TABLE 5.1
Various knitting methods with their construction and compression application therapy

Knitting method	Knitting machine	Knitting construction	Isotropy in deformation	Applications
Weft knitting	Single jersey Double jersey	Elastic yarns are knitted or inlaid or plating to knit jersey or double jersey, fleece, pique	Elastic deformation in the wale and course direction	Sportswear; medical compression garments
	Rib interlock	Bare or covered elastic yarns are inlaid or plating	Elastic deformation in the wale direction	
	Flat knitting	i. Covered elastic yarns are knitted or inlaid or plating ii. Space fabric	Elastic deformation in the wale and course direction	
Warp knitting	Tricot	Elastic yarns are knitted on the back bar with 1-0/1-2/tricot stitch to knit fabrics with plain jersey and mesh	Elastic deformation in the wale and course direction	Swimwear; tight-fitting garments
	Raschel	Elastic yarns are as laid-in stitch to produce power-net construction	Elastic deformation in the course direction	Swimwear; tight-fitting garments;
	Double-needle-bar raschel	i. Elastic yarns are in the surface bar to knit space fabric ii. Circular elastic fabric	Elastic deformation mainly in the course direction	support garments

Source: [4] (Reprinted under Creative Commons licence)

measures the pressure variation when a patient moves from the supine to the standing position. The latter one measures the pressure pulsation during the dynamic wear of the compression garment. Proper analysis and optimization of the material stiffness properties will help in balancing the effectiveness and wear comfort of the compression garments. The second most important parameter is the elasticity of the material. It is the ability of the textile material to return to its original state after the removal of stress. This behaviour of the material can be expressed by Hooke's law. However, the textile materials cannot obey the Hooke's law due to their anelastic behaviour. This phenomenon represents the dynamic pressure performance of the material when it is worn for a longer duration. During sports and other activities, the body imparts continuous stretch and relaxation on the garment affects the compression behaviour of the material. The lower hysteresis value of the elastic material represents the durable pressure application in a compression garment [6]. The elastane fabric or compression garments are classified based on the applications in the medical field. Various medical applications of compression garments are given in Figure 5.2.

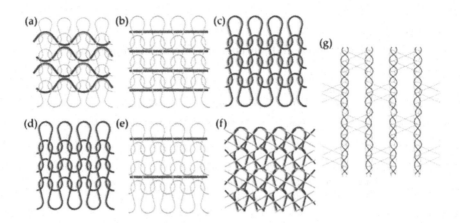

FIGURE 5.1   Typical knitting construction for elastic fabrics (dark line is elastic yarn and the light coloured line is ground yarn): (a) Elastic yarns are as a laid-in stitch in jersey. (b) Elastic yarns are as weft inlay stitch in jersey. (c) Elastic yarns are as plating stitch in jerseys. (d) Elastic yarns are as plating stitch in rib. (e) Elastic yarns are as weft inlay stitch in jersey. (f) Elastic yarns are knitted on the back bar with 1–0/1–2/tricot stitch. (g) Elastic yarns are a laid-in stitch to produce power-net construction [4] (Reprinted under Creative Commons licence).

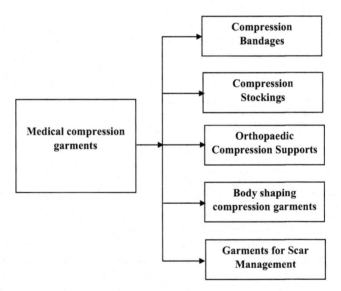

FIGURE 5.2   Classification of compression garments used in the medical field (Authors' own illustration).

5.2   COMPRESSION STOCKINGS

Compression garments used in the bottom portion of the human body particularly in the leg portion are known as compression stockings (Figure 5.3). These garments are engineered to provide different compression on the different parts of the leg portion based on the requirements. In common, the higher pressure applied in the bottom or

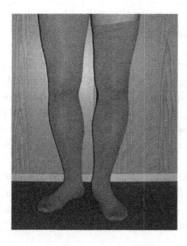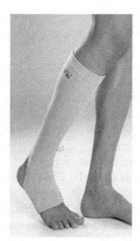

FIGURE 5.3 Compression stockings in medical applications (Reprinted under Creative Commons licence).

ankle portion of the leg and the compression pressure steadily reduces towards the upward direction. In general, the stockings are made of rubber or elastane yarn in the transverse direction that is responsible for compression behaviour. Sometimes, the weft yarn can be of core elastic and covered with other fibres like cotton on the outer shell for better aesthetic and comfort properties. Concerning the performance of the stockings, there are several standards from nations like Dutch (RAL-GZ 387), French (NFG 30–102B IFTH), and Britten (BS-7505) developed to standardize the manufacturing process and ensure proper pressure in application. To develop such kinds of stockings with differential pressure, it is essential to construct the garment with engineering fabric properties [7]. There are three different types of compression stockings in the market as listed below [8].

i. **Graduated compression stockings:** These include the garments that provide higher compression pressure in the ankle region of the stockings and the pressure level decreases towards the up. The graduated stockings are prescribed mostly for medical application. The types of stockings that end below the knee are commonly used for issues like peripheral edema or lower leg swelling. The stockings that are extended to the waist are commonly used to reduce the cooling of blood in the leg.

ii. **Anti-embolism stockings:** These are the type of stockings that are very similar to the graduated compression stockings, but the stockings differ in their pressure level which is highly based on the requirements of the patients. These stockings are mostly used for people who are not mobile.

iii. **Nonmedical support hosiery:** This class of compression stockings is a commercially available and low-pressure range product. The purchase of these category stockings does not require a prescription from a doctor. They are commonly used for relief from leg aching and tiredness. These types of stockings deliver an equal amount of pressure throughout the applied area.

However, based on the compression level, the compression stockings are also classified under four different categories as follows [9]:

a. **Commercial stockings:** Commercial stockings are lower pressure level garments and they can be purchased commercially without a medical prescription. The pressure range of this garment is ranging from 15–20 mmHg.
b. **Medical Class I:** The class I category represents the garment pressure range of 20–30 mmHg. This type of stockings is medically prescribed and used for issues like leg swelling, spider veins, varicose veins, travel, sports, and after certain surgeries.
c. **Medical Class II:** Class II medical compression garments are used to provide pressure in the range of 30–40 mmHg. This class is recommended during the issues like blood clotting, deep vein thrombosis (DVT), and patients with lymphedema are also recommended to use the 30–40 mmHg level of compression.
d. **Medical Class III:** This category of the stockings provides a higher pressure range of 40–50 mmHg. These stockings should be used only under the recommendation of medical practitioners.

5.2.1 Mechanism of compression stockings

The graduated compression garment provides higher pressure at the ankle region and the compression gradually reduces while it goes up. This gradual change in the pressure increases the blood flow upward to the heart instead of the downward refluxing to the foot. Application of pressure in the leg portion reduces the diameters of the major veins and so it increases the velocity of the blood flow in the veins. Figure 5.4 represents the mechanism of compression stockings.

The use of a compression garment can reverse a few issues in the leg portions like venous hypertension, augment skeletal-muscle pump, smoothen the progress of venous return, and improve the blood flow. There are unproven results that support that the use of compression garments develops complex physiological and biochemical effects, improved tissue oxygenation after the use of compression stockings [10].

5.2.2 Effect of elastane and its fabric properties on compression stockings

Though several advantages occurred due to the use of compression stocking, the ability of the compression mainly depends upon the elastane or spandex yarn used in the compression garment. Hence, the changes in the elastane properties, types, and way incorporated inside the garment plays a vital role in the compression ability of the garment. Elastic recovery is one of the crucial parameters that determine the performance of the garment. In general, when an elastic material is stretched to some extent that is lower than its breaking strength, it immediately returns to its original state. The recoverability of the compression fabric strongly depends on the compression force provided and the length of the time permitted to recovery. In order to maintain durability, the residual extension should be maintained as low as possible to avoid fatigue [11].

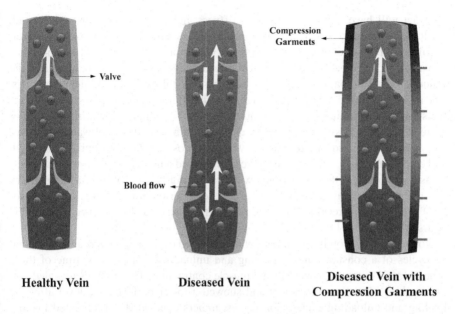

Healthy Vein   **Diseased Vein**   **Diseased Vein with Compression Garments**

FIGURE 5.4   Effect of compression stockings on the leg portion (Authors' illustration).

In a study, the relationship between the compression behaviour of the stockings and the linear density of the elastic yarn was compared. The findings revealed a direct positive correlation between the evaluated parameters. An increment in the elastic yarn linear density used in the compression garment has directly increased the compression effect. On the measurement with stockings made of different linear densities from low to high, researchers noted an increment from 16 to 19 mm Hg at the ankle, 13–17 mm Hg in the calf region. They also investigated the increment in the linear density of the inlay elastic yarn on the compression effect. For an increment of linear density from 133 dtex to 330 dtex, they found an overall increment of the pressure from 15 mmHg to 23 mmHg at the ankle region and a similar increment in the calf region also. The results were also carried out on the influence of the draft value of the yarn on the compression behaviour of the garment and found no significant changes in it [12].

Burak Sari and Nida Oglakcıoglu evaluated the influence of fabric production parameters on the compression properties of the stockings. In their study, they reported that the fabric thickness has a significant impact on the compression ability of the stockings. They noted an increment in the compression along with an increment in the fabric thickness. They also measured the influence of elastane linear density and reported the impact is significant. The increment in yarn diameter or linear density significantly increases the compression effect of the garment. They analysed the combined effect of these two parameters. The increment in linear density increases the fabric thickness and so the compression of the garment. It is noted that the process parameters like stitch density, structure compactness, and yarn linear densities used to have a significant impact on the compression ability of the stockings. Further, it is noted that the garment with higher structural thickness and compact will have higher weight and stiffness and so when the garment is used, the force required to stretch will be applied back as a resilience force in the wearer. Similarly, it is also reported that the traverse elastic property of the textile fabric provides a negative impact on the compression garments [13]. Wang et al. measured the influence of fabric parameters on the compression ability of the medical stockings. The researchers used nylon/spandex fabric with different grams per square metre for the study. Their findings revealed that the fabric with a higher breaking strength of more than 200 N possessed higher extensibility in both course and wale directions. The findings reported that the fabrics were also possessed higher bursting strength and extension values. The fatigue test results showed a higher (95%) immediate compression recovery in the nylon/spandex fabric. Similarly, 98% of the elastic recovery was noted in these fabrics for a 24-hour extended period of relaxation. A very lower (2%) percentage of residual loss was noted after three weeks of service and a few hours of relaxation [14].

To measure the compression loss by the repeated cyclic load, a researcher applied 20 cycles of a constant rate of loading and unloading with a dwell time of three minutes after every cycle to replicate muscle contraction. The findings revealed that the stockings lost their efficiency and showed a lower performance level after the loading and unloading cycles. So, the researchers concluded that repeated loading will reduce the tensile strength values and so the compression ability value of the stockings. They also found the impact of dwell time on the tensile strength values of the compression stockings in regular applications [15]. To measure the influence of mechanical properties on the compression properties, researchers used the Kawabata system. The findings of the analysis of the mechanical properties such as tensile, bending, and shearing and surface properties of various compression stockings samples indicated that only the tensile parameters showed a strong correlation with compression pressure, whereas the shearing (G, 2HG), bending (B) and weight (W) properties showed a positive but lower correlation than the tensile properties [16]. In a comparative study, flat- and round-knit fabric applications in compression stockings were evaluated. In findings, the researcher reported after a systematic literature review that as of now no scientific or clinical studies were found to compare the effectiveness of the flat and round knit compression stockings. Further, the significance of different knitting techniques was not technically estimated through any clinical study [17].

Homa et al. studied the effect of stitch length on the compression behaviour of the plain and interlock knitted fabric. In plain fabric, the lower stitch length showed a higher deformation due to the insufficient space in between the yarn in the structure. Hence, the applied load fell on the fibres in the yarn instead of the fabric, and this creates a time-dependent deformation in the fabric structure. But in the case of the interlock structure, there is enough space available between the yarn structures. Hence, the applied load did not cause any structural deformation like plain structure. The researcher also evaluated the same effect after washing and repeated usage. Though the laundry showed some initial reduction in the pressure decrease after the seventh wash, plain fabric showed a greater reduction in the pressure. In the case of interlock fabric, there was a pressure reduction up to the fifth wash, and after that constant pressure was noted [18]. To evaluate the influence of the mechanical and surface properties of the fabric on the pressure and comfort aspects of the compression hosiery, fabrics were studied by the researcher. The mechanical properties of the material were estimated using the Kawabata Evaluation system and pressure measurement was done using a skin pressure measuring system. The compression stockings fabric with higher compression pressure and lower compression pressure were used in this study to evaluate the mechanical properties of the material. The findings of the results revealed that the mechanical properties showed a significant influence on pressure levels developed by the system. The results revealed that tensile energy, tensile strain, shearing, stiffness, and bending rigidity are key mechanical material indices that have a direct influence on skin pressure. Out of the selected materials, the soft or low-pressure compression garments showed a higher correlation with the fabric's mechanical properties. Whereas, higher pressure garments showed a lower correlation with mechanical properties. The higher correlation of the lower pressure garments mainly correlated to the smoother surface, higher elasticity, and better dimensional property. Higher compression fabrics were rougher, stiffer, and less extensible, but they had better dimensional stability. Kawabata analysis revealed that the mechanical properties were significantly different for the higher-pressure and lower-pressure garments even at the time of similar pressure applications. The findings of the research confirmed that the lower surface roughness property and proper shape or design ensures the proper pressure application on the body [19].

5.3   COMPRESSION BANDAGES

Compression bandages are stretchable fabric-based bandages that are commonly applied by wrapping around the sprain or strain. After application, the pressure imparted by the bandage helps in reducing the pain or swelling in the particular parts of a body. Several research reports mentioned that compression bandages help in increasing the healing rate of venous ulceration in the leg as shown in Figure 5.5. The fundamental mechanism is that the application of the compression bandage increases the venous blood velocity and reduces the superficial capillary and venous pressure [20]. Among all available compression bandages, multilayer compression bandages are more efficient than single-layer bandages. Compression bandages are divided into three main categories [21]:

- **Type 1. Lightweight conforming-stretch bandages:** These bandages are used for simple dressing functions. The main objectives of these bandages are to confirm the body contours well and to have the stability to withstand the body movements.
- **Type 2. Short-stretch bandages:** These bandages are generally made of 100% cotton fabric. Hence these bandages are a non-elastic item, which has the minimal stretchability of the fabric. Short stretch bandages are used for rigid casing around the limb [22].
- **Type 3. Long-stretch bandages:** These types of bandages are also called extensible or elastic bandages. These bandages are commonly manufactured by synthetic fibres using knitted or woven structures. These bandages can be stretched up to 120% as elastomeric yarns included in its structure [22].

Further, the elastic or long stretch bandages are classified into four different categories as reported below based on the level of compression which imparts on the body part. The types of bandages are mainly used in the applications of venous disorders of the lower limb.

- **Type 3a. Light-compression bandages:** Designed to provide a lower level of pressure on the body parts. The pressure value is limited to the 14–17 mm Hg in the ankle region. This type of fabric is commonly used in the beginning stage of varices and varicose formed during pregnancy by providing a low level of pressure.
- **Type 3b. Moderate-compression bandages:** This kind of bandage can exert a pressure of 18–24 mmHg on the ankle at every dimension.
- **Type 3c. High-compression bandages:** The pressure range varies from 25–35 mmHg. Generally, it is used to apply a high level of compression in the event of gross varices, post-thrombotic venous insufficiency, and the management of leg ulcers.
- **Type 3d. Extra-high-performance compression bandages:** It can create with a higher amount of pressure level of above 50 mmHg [23].

5.3.1 STRESS RELAXATION – A COMMON ISSUE IN TEXTILE MATERIAL

In the compression bandages, the stress applied on the limb is generally developed by the application pressure. The ability of the bandage to exert pressure on the wrapped place is mainly based on the type of textile structure and fibre materials used in the fabric. Based on the fibre type used, its viscoelastic behaviour develops internal extension and so the stress applied to the limb reduces due to the relaxation of the fibres and yarns in the structure [24]. This stress relaxation is mainly due to the release of internal stress over a period of time with the same level of strain. This phenomenon significantly influences the compression pressure of the bandages. The stress relaxation behaviour of the textile material mainly depends upon the fibre type, yarn structure, and fabric structure used in the bandage (woven or knitted). Research reports mentioned that the different fibres have different stress relaxation behaviour under different stress conditions [24, 25]. The maximum amount of stress relaxation is noted with cotton and viscose fibres than synthetic fibres like polyester and nylon [26]. Understanding the relaxation behaviour of the textile material used in the

FIGURE 5.5 Compression bandages for medical applications (Reprinted under Creative Commons licence).

compression bandage is one of the essential requirements to maintain the constant pressure over the time of application.

In a study, researchers evaluated the stress relaxation performance of the cotton, viscose, polyester, cotton-Lycra, and polyester-Lycra core-spun yarns. The results reported that the tightness factor of the fabric played an important role in the pressure reduction in the compression bandages. They also reported that the lower pressure drop in the compression bandages with a higher tightness factor showed a lower pressure reduction than the lower tightness factor of the same yarn. The main reason behind the reduction in pressure is the number of yarns in the structure. As the higher tightness factor shares more yarns in its structure to distribute the stress applied, the overall stress relaxation reduces and so the pressure drops from the bandage, over a lower tightness factor fabric. Similarly, the case of lycra core-spun yarns also showed a lower stress reduction compared to the 100% cotton and viscose fabrics after eight hours of trial. It is due to the elastic nature of the core fibre and the higher tightness factor of these structures over the plain knit structure. First, the elastic yarns can maintain internal stress for a longer time, and secondly, the higher compact structure of the core-spun yarn fabric exerts lower stress on the individual yarns in the structure [27].

In another study, researchers evaluated the effect of fabric tensile strength and porosity on the compression behaviour of the compression bandages. The study evaluated the compression pressure at static and dynamic conditions. The results reported that the extension of the bandage increased the porosity of the fabric and the weave angle of the yarn present in it. They reported that the pressure variation of theoretical evaluation using Laplace's equation reaches a maximum of 20% at the extended state. By analysing the 100% cotton, cotton/polyamide/polyurethane, and viscose/polyamide bandages, the results reported that the pressure applied on the leg is directly proportional to the bandage extension and its porosity. This parameter remains common for all the samples tested in the study. To achieve a normal pressure of 10 N load, there is a need for a 60% extension on the bandage material while 100% cotton is used. But at the same time, a higher extension is required with cotton/polyamide/polyurethane and viscose/polyamide bandages as reported in Figure 5.6 [28].

The pressure-creating ability for compression bandages is mainly based on the fibre, structure, and type of yarns used in the structure. Concerning the pressure analysis, researchers evaluated the pressure generation in different commercial bandages. The results of the study showed that the different material bandages used under the same pressure category, the pressure development varies a lot on the application. The structural aspects of the garment decide the tensile strength, stretch, and relaxation behaviour. In their results, the bandages with coarser elastic yarn showed a higher recovery and lower structural disintegration during the usage. Similarly, the bandages with high stretch elastic yarns and tighter structure provided a strong and durable fix on the limbs of the wearer and were suitable for longer usage or repeated usage [29]. Other researchers compared the four knit structures and bi-stretch woven fabrics with the same areal density for comfort, pressure, and mechanical properties after different laundry cycles. The study used single jersey, single Lacoste, plain pique, and honeycomb nit structures and 1/1 plain, 2/1 twill, 3/1 twill, and 4/1 twill

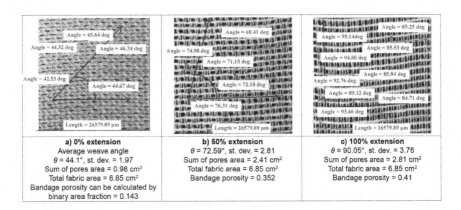

a) 0% extension	b) 50% extension	c) 100% extension
Average weave angle	θ = 72.59°, st. dev. = 2.81	θ = 90.05°, st. dev. = 3.76
θ = 44.1°, st. dev. = 1.97	Sum of pores area = 2.41 cm²	Sum of pores area = 2.81 cm²
Sum of pores area = 0.98 cm²	Total fabric area = 6.85 cm²	Total fabric area = 6.85 cm²
Total fabric area = 6.85 cm²	Bandage porosity = 0.352	Bandage porosity = 0.41
Bandage porosity can be calculated by		
binary area fraction = 0.143		

FIGURE 5.6   Relationship between extension, weave angle, and porosity for cotton bandage [28] (Reprinted under Creative Commons licence).

woven structures for the study using core-spun elastane yarn. The researcher measured the pressure development, air permeability, and stretchability of the selected fabric after 5, 10, and 15 washes. The results revealed that the pressure performance of woven fabric was better than the knitted fabric. In woven fabrics, before laundry, plain fabric possessed higher pressure generation capacity due to its structural compactness. After 15 laundries, a pressure drop percentage of 17.3, 16.6, 14.1, and 17.1 were noted for the woven structures 1/1 plain, 2/1 twill, 3/1 twill, and 4/1 twill weave, respectively, than the unwashed fabric. In the case of the knit structures, after 15 washes, a pressure reduction percentage is 38.2, 34.3, 34.1, and 32.5 in single jersey, single lacoste, plain pique, and honeycomb structures, respectively. Though all the garments showed better retention of pressure after 5 washes, the woven fabric possessed lower pressure loss due to its structural compactness after 15 washes. However, in the case of comfort, air permeability and stretchability parameters were noted higher with knit fabric than woven fabric. Hence, the researcher suggested considering these parameters while designing and selecting materials for compression therapy applications [30].

Thermal comfort properties of the compression bandages are important while are using the compression bandages for a longer duration. Abdelhamid et al. measured the thermal comfort behaviour of four different commercial compression bandages using the thermal foot model, to measure thermal resistance and water vapour resistance of the compression bandages. The results of the research represented that the application of extension in the compression bandages significantly reduces the thermal resistance of the bandages at 10%–40% extension, due to the reduction in the thickness. But at the same time, the researcher noted an increase in thermal resistance at the extension range of 40%–60%. However, after 80% extension, the thermal resistance drops again due to the increment in the compression values and reduction in bandage thickness. They also noted the increment in the applied tension from 0.5 to 10 N, significantly reduced the thermal resistance of the material. It is also noted that while increasing the number of layers on the limb, the thickness of the layers reduced and so did the thermal resistance of the bandage. In the case of water vapour resistance, the researchers found a similar effect of thermal resistance. An increment in the extension reduced the water vapour resistance and also air permeability due to the reduction in the thickness of the compression bandage fabric. In total, the air permeability of the compression bandages of all fabric increases when the bandage extension increased from 0 to 100% [31].

In a clinical study, medical practitioners evaluated the interfacial pressures generated by elastic and non-elastic bandages using in-vitro analysis. The researchers used a randomized trial by providing both the bandages for all participants. In the study, the researcher used 100% cotton compression bandages as non-elastic material and cotton/lycra/polyamide bandages as an elastic material. After the wear trial, the findings of the study revealed that the interface pressure varied as the stiffness of the material differs. Though different bandages were taken with various pressure differences noted in the different regions of the bandages for the study, the common pressure difference between the bandage types wherein the range of 13 mmHg. The inelastic bandages had a higher stiffness (7.3 mmHg) than the elastic ones. During the activities, the pressure inside the bandages varied in the range of 15.5 mm Hg. The findings confirmed that compared to the elastic bandages, the inelastic bandages showed a higher compression during the resting and activities. The results confirmed the important role of fabric stiffness on the compressibility of the compression bandages [32].

5.4 ORTHOPAEDIC SUPPORTS

The garments which are used in the place of knee, elbow, and ankle joint traumas are generally known as Orthopaedic supports. Based on the severity of the trauma, different types of supports were used in the medical field. The main motive of orthopaedic support is applying pressure through elastic garments on the limbs of the body. The typical Orthopaedic support garments are provided in Figure 5.7.

Knitted orthopaedic supports can be used in different parts of the bodice like knee braces, wrist braces, ankle braces, shoulder braces, elbow braces, calf, lumbar, and back supports. The Orthopaedic supports are mostly customized for the subject based on the requirements and the prescription of the medical practitioner. However, the compression requirements are not standardized to date. The use of orthopaedic supports increases the enhanced motor skills, body strength, and/or provides support to paraplegic patients, nursing women, and patients with motor disabilities. The orthopaedic supports are broadly classified into three different types [3]:

• Preventive supports
• Functional supports
• Postoperative/ rehabilitative supports

As reported in the previous sections, the supports are also available in pressure variation, and the supports are also grouped under the amount of pressure it exerts. Light-compression orthopaedic supports are termed as class 1 (15–21 mmHg), class 2 (23–32 mmHg), class 3 (34–46 mmHg) and class 4 (>49 mmHg) [33]. The orthopaedic support evaluates the biomechanical condition of muscle and provides constant

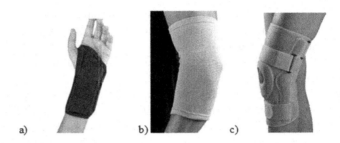

FIGURE 5.7 Orthopaedic supports at (a) wrist; (b) elbow; (c) knee (Reprinted under Creative Commons licence).

pressure on the spot that requires treatment. At the same time, it is important to retain the geometrical shape and compression ability of the material for a longer duration even after repeated use and laundry. The medical compressive orthopaedic supports must be apt for the patient's limb size based on the application area of the body [33]. The compression ability and the durability of the supports are mainly based on the type of yarn, yarn linear density, type of elastane yarn, the production process, and type of machine used in production. Out of all these parameters, the elastic properties of the yarn or elastane thread majorly influences the fabric properties. Most of the time, the orthopaedic support material also consists of neoprene along with the knitted fabric [34]. Foams, unilateral or bilateral bars with hinges, and straps are also used in the support to hold the weight of the body. The main role of elastane in the fabric is to provide compression on the affected area and also provide extensibility to the braces during the movement of the body.

Researchers evaluated the effect of washing cycles on the compression properties of the knitted orthopaedic support. Upon laundry, the fabric shrinks, and the loop density increases in the fabric. However, the researcher reported that the shrinkage was reduced with repeated laundry. The results revealed that the knitted orthopaedic support shrunk significantly after the washing and drying cycle. The shrinkage changed the orthopaedic support and made as thicker and tighter which is due to the changes in the loop density of the fabric. The researcher found that the shrinkage value increases with an increase in the duration of laundry. This structural change showed significant changes in the pressure imparted on the body. After the first wash, the orthopaedic support fabric showed an 11% reduction in the surface, and the pressure increased by 3.6% on the limb. However, successive washing drying did not show much influence on the pressure [33].

In a study, orthopaedic support fabrics were produced using different inlay densities of the elastic yarn using a double-needle flat-bed rib knitting machine. The findings of the research reported that the inclusion of inlay yarn in the knit structure had a significant amount of influence on the compression of the orthopaedic support fabric. The results mentioned that the changes in the inlay yarn linear densities did not have any impact on the compression pressure of the orthopaedic support fabric. Similarly, the role of cover yarn is also noted with insignificant influence. But it is noted that the number of insertions of inlay yarn has a direct influence on the compression pressure generated. An increment of 21%–25% in the compression pressure was noted with the fabric developed from the inlay yarn inserted in every course of the knit than the inlay yarn inserted in every second course of the feed. It is also observed that the reduction in inlay yarn densities from two inlay yarn in every second course to one inlay yarn in every second course reduced the developed compression pressure of 6%–13%. The researchers also investigated the effect of rigid elements inside the orthopaedic support. They reported that a strong dependency on the compression developed and rigid material used. An increment in the area of the rigid support directly increases the compression pressure of the garment. By this research, they estimated the role of rigid material and the effect of elastic yarn inlay densities on orthopaedic support compression pressure [35].

In a later study, the researcher studied the role of inserted materials in the orthopaedic supports. The inserted materials significantly affect the elasticity of the orthopaedic supports. However, it is important to understand that the area of the insert rigid material has a strong influence on the compression effect generated by the orthopaedic supports. The research results mentioned an increment of 15% compression in the orthopaedic supports by including a rigid element up to 8% of the total area of the support. In the case of orthopaedic supports with rigid elements that cover 25% area, an increment in compression noted up to 17% at 10% elongation and 24% increment with a 20% extension. The study also reported that the use of a rigid element of less than 3% of the total area of the Orthopaedic supports did not have any influence on the compression pressure [36]. Concerning the analysis of mechanical behaviour, research works performed with knitted fabric made up of two different types of core-covered elastic yarns along with a biocompatible silicon pad in it. The stretchability of the samples was analysed by stretching the sample to a maximum and kept stable for 120 seconds. The results reported that the normal common method of stretch evaluation cannot be used for the different components of the body. It is reported that the knee supports and ankle supports should be stretched for 20% to achieve targeted tensile force. However, in the case of the wrist, it required only 115% extension to exert the same level of pressure. The researchers mentioned that based on the perimeter of the limb, the perimeter of the compression product should also be modified [37].

Other researchers used the spacer fabric for the application of knee braces, a type of orthopaedic support, for their study. They developed three weft knitted spacer fabric and one warp knitted fabric. They compared the results with commercial weft knitted and neoprene-based garments. Spacer fabrics are special fabrics made of three-dimensional structure. The fabric consists of face and backside fabrics which are interconnected by monofilament yarns like polyester, polyamide, etc. The fabrics were developed in all kinds of knitting machines like flat, circular weft, and warp knitting machines. The fabric provides a bulky structure with breathability, comfort, durability, and resilience under load. The fabrics also offer a higher stretch in all directions and also, they are dimensionally stable [38]. The researcher evaluated the mechanical and comfort properties of the developed spacer fabric and commercial fabrics. The results reported that the knitted fabric showed a higher mechanical property than the neoprene fabrics. The spacer fabric also possessed a similar value to the knitted fabric. They reported that the knee braces will be used for a longer time as per medical advice; hence, it is necessary to measure the comfort properties of the developed orthopaedic supports. Out of all the structures, neoprene-based products offer a wet and uncomfortable to the wearer irrespective of the type of activity performed. In the case of water vapour permeability, absorption, and wicking test results, spacer fabric outperformed, compared to all the other samples and commercial products. The results suggested that compared to the commercial products, knitted spacer fabrics were more suitable for orthopaedic supports [32].

Diana Ališauskienė et al. evaluated the inlay yarn linear density and insertion density on the compression ability of the knitted orthopaedic supports. The researchers used a flatbed knitting machine and they varied the inlay yarn insertion and linear density. The results of the study reported that the increment in the elastic yarn insertion density increases the pressure in the garment significantly. The higher pressure noted in the support produced with higher insertion density. Approximately 50%–127% of the pressure increment was noted in the four different structures tested. The insertion of additional inlay yarn in every second course did not show any significant impact on the pressure. Instead, the structure becomes stiffer and rougher than the previous one, a pressure increment of 7.8%–16% noted in this case. The research reported that the inlay yarn insertion density increases the knitted orthopaedic support's pressure generation capacity exponentially. Concerning the linear density of the elastic yarn, the results showed a linear relationship between the pressure value developed and inlay yarn linear density only at the lower extension level (up to 10%). For samples with lower insertion density, even after six times increment in the linear density did not show a significant increment in the pressure level. But at the same time, the researcher noted that the sample with higher insertion densities showed a significant change in pressure when the linear density of the inlay yarn increases twice. The researchers also found that there is no influence of cover material properties and densities on the compression of the orthopaedic supports [39].

5.5 COMPRESSION GARMENT FOR BURN SCAR

Burn scar management is also performed with pressure therapy, which is also known as compression therapy. This treatment is one of the essential processes in the patient burn management process. The application of elastic bandages and compression garments in this sector is also one of the notable applications of elastic fabrics. The application of the compression by fabric reduces the development of the scar by reducing the collagen production in the burn area. The pressure application technically protects the skin from being fragile. The pressure application promotes the circulation of damaged tissues. Further, compression therapy reduces itching, pain, and thickness of the scars [40]. In burn scars, hypertrophic and keloid scars are difficult to manage as the mechanism of healing is not known. Technically, hypertrophic scars appear as a result of the wound healing process. These scars happen within a few weeks of burn and they are normally red, thick, itchy, and painful. In the case of keloids, the scars use to grow larger than the wound area. This cannot be predicted and it's common for dark skin people from countries like Africa and Asia [41]. In burn wound management, the prevention of hypertrophic scars is one of the important issue. Compression therapy treatment was traditionally used in hypertrophic scar treatment from the early centuries. Either by applying mechanical loading or by compression up to 6–90 mmHg used to recover from several medical and pathologic conditions. In the area of soft tissue, a pressure range of 9–33 mmHg is used, and in the bone, a prominent area, the pressure up to 47–90 mmHg is used. However, the use of pressure or elastic garments in hypertrophic burn scars was in practice only after the 1970s [42]. Due to the non-invasiveness, compression therapy became one of the desirable treatments for hypertrophic burn scars. Pressure therapy is used in the burn scar treatment based on three main reasons:

i. The use of a compression garment restricts the blood flow in the wound area due to the applied pressure [43].
ii. The constant pressure application in the wound or scar controls collagen synthesis as a result of reduced or controlled oxygen supply, blood supply, and nutrients. By so, the pressure application speeds up the maturation process [44].
iii. The pressure re-aligns the collagen bundles in the scars and reduces the hydration along with neovascularization and extracellular matrix [45].

Thus, the pressure application process reduces the generation of collagen and reduces the thickness, and softens the scar tissue area. However, it is essential to mention that the short application of pressure may not be suitable to cure the scar. It is important to maintain the prescribed pressure for a long time or as mentioned by the medical practitioners. Though several studies were performed in the medical field to evaluate the performance of the compression garments in burn scars, no studies were carried out in the field of textiles. It is a kind of area where the benefits of pressure garments are not fully understood or explored. The previous research results were differing on several factors in developing a positive impact on the burn scar.

Jayne Y. Kim et al. evaluated the effectiveness of the compression garment on the burn scar. The researchers developed a burn wound model on a female Red Duroc pig and after 28 days of post-burn wounds, the compression bandages were applied and the effectiveness in scar management was analysed after 28, 56, and 72 days. In this study, the pressure garments are used with a pressure of 10 mmHg and compared with a controlled wound that is covered with a normal fabric without pressure. The results of the study reported a significant difference in the scar contraction of the control wound and pressure applied wound scar. This is mainly due to the compression force applied to the wound spot. The pressure applied increases the phosphorylation of focal adhesion kinase and regulated the smooth muscle actin and collagen production. Hence, the applied pressure reduces the differentiation of fibroblasts to myofibroblasts difference. This, ultimately decreases scar contraction and collagen deposition. Hence, the findings of the study confirmed that after 72 days of observation, the wound contraction in the pressure garment's used portion was 82.7% and, in the control, it was 64.6%. Compared to the control wound, pressure garments aided a better scar contraction in burn scar treatment. Further, the researchers also noted a reduction in skin hardness and a 1.3-time increment in skin strength [46].

Though research works to support the pressure applied as a method of burn scar, no standard pressure measurement/ methods available so far. The administrated pressure level is one of the important controlling factors. Hence, without a standard and reliable method, the results of the studies obtained from pressure therapy were not reliable. Though recent studies suggested the advantage of compression garment usage in scar management, the application is limited to those who are with moderate or severe scarring [47]. In a review, Atiyeh et al. analysed the usage of pressure therapy and reported that the evidence of complete recovery was not clear. This is mainly due to the multiple variables like the degree of hypertrophic scarring, inability to quantify pressure applied to scars, patient noncompliance to strict pressure therapy time schedules associated with compression garment usage [48]. In a most recent study, the researcher evaluated the management of burn scar with pressure therapy and compared it with topical silicone gel and both combined applications of both silicon and pressure garments. In this in-vitro analysis, the researcher used 153 subjects out of it they applied silicon for 51 participants, pressure garments for 49, and 53 participants with both. Six months after a burn wound, the combined treatment of pressure garment and silicone treatment did not have any effect compared to other methods applied. With combined treatment, the scar was thickened than the other groups. The researcher did not notice any difference between the silicon and pressure groups. However, the researchers noted some adverse effects of pressure application on scars [49]. A similar review was also performed by other researchers. They compared the results of various clinical trials performed based on compression garment designs and pressure levels. However, the article concludes that the effect of garment pressure on scar management was clinically not proven well. The literature did not show any scientific method of pressure measurement to confirm the same. Though few studies demonstrated the benefits of pressure garments, those studies did not consider all the influencing parameters. Additional information regarding pressure therapy's effects on scar management can be found under these review articles [48, 50]. The following points represent the limitation of pressure garments in scar management as reported by Atiyeh et al. [48].

- The major disadvantage of pressure therapy or the use of compression garments is the unknown interfacial pressure between the garment and the body. The lack of scientific methods to measure the pressure made the exiting research results are invalid.
- Previous research reports confirmed the pressure on the bone prominent area is lower and the tissue area is higher than the predicted pressure of the garment.
- Similarly, the pressure on veins was noted high than the pressure on the limbs.
- Repeated usage of garments reduces the pressure of the garment. Hence, it requires frequent replacement. Approximately after one-month usage, the pressure garment losses its 50% pressure.
- Garments that exert higher pressure lose their ability on washing faster than the garment with lower pressure generation.

5.6 COMPRESSION GARMENTS AS SHAPEWEAR

Shapewear are the garments generally used to smoothen the body shape. It is usually worn by women as a way to improve their body shape. The shapewear garments are underwear, used for adjusting the wearer's body shape temporarily and producing a fashionable slim look [51]. Generally, shapewear is categorized as lingerie in most of the stores; however, it is the foundation garment with a specified level of pressure application. Many shapewear garments are commercialized and there are very few research works performed in the shapewear garments and fabrics. Shapewear can be of corsets, girdles, or bodysuits but they are potentially changing the shape of the body by providing a perfect silhouette. Shapewear is generally made of nylon fibre along with elastane. However, they are available with different fibre combinations to impart comfort to the wearer. Manufacturers developed lighter thick materials for better breathability for a warmer climate and strong materials for cold climates. Shapewear garments are generally classified into three broad categories as provided in Figure 5.8.

5.6.1 COMMERCIAL BODY SHAPER

Commercial body shapers or shapewear are the elastomeric garments used for aesthetic purposes. These garments are available in different shapes based on their applications as camisole, slip, bodysuit, high waist brief, thigh slimmer, waist shapers, etc. [52].

5.6.2 POSTPARTUM BODY SHAPER

These are the shapewear used in the time of post-maternity usage. It helps the body to recover from the pregnancy. It aids in recovery by assisting with abdominal support, avoiding itching, etc. It helps in achieving the pre-pregnancy body shapes and posture [53].

5.6.3 POST-SURGERY BODY SHAPER

These are the common body shaper similar to the postpartum body shaper. As the previous one focuses on pregnancy, this types aids in recovering from common

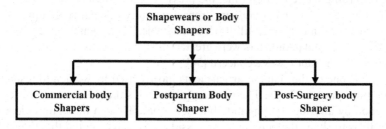

FIGURE 5.8 Classification of shapewear garments (Reprinted under Creative Commons licence).

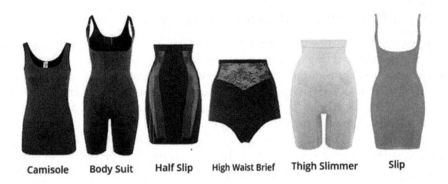

Camisole   Body Suit   Half Slip   High Waist Brief   Thigh Slimmer   Slip

FIGURE 5.9   Elastane knitted fabric as body shapewear.

surgeries like abdominal work, a tummy tuck, liposuction, breast surgery, etc., as provided in Figure 5.9.

In shapewear, development and efficiency analysis are the two major areas where much researches are required. Based on the application requirements, manufacturers used different fibre combinations with elastic yarn. No detailed research works were performed in the body shaper sector as the general property analysis like compression bandages and compression stockings are common for these garments also. However, the majority of the literature focuses on pressure measurement and optimizations. Many researchers took the challenge of measuring and developing optimum pressure levels for various shapewear. Few of those research works were detailed in this section to establish the role of elastane fibre and its importance. Researchers examined the effect of the mechanical properties of fabric on compression behaviour. In this study, the researcher used cotton/spandex and polyester/spandex elastane knitted fabric on a cylinder model to evaluate pressure development. The fabric extension on both warp and weft directions was measured using pressure sensors on the cylindrical model. The effect of elastic modulus, elongation percentage, and relaxation time on the pressure development behaviour was detailed. The results of the analysis revealed that higher elastic modulus provides higher compression pressure on the human body at the same elongation of the lower elastic modulus sample. In the case of elongation, the pressure applied was noted higher with the fabric with higher elongation. Relaxation time is another crucial factor in the case of compression fabrics. Concerning the relaxation time, the pressure reduction followed the first-order exponential trend. The major pressure drops were noted in the first 30 minutes of the relaxation and then continued till it becomes stable. The researcher mentioned that all the three selected parameters were greater influencing factors when the compression ability of the fabric was a concern [54].

Girdle is one of the shapewear garments mostly used by women to reshape the bottom portion of the body. The use of a girdle helps the women to look aesthetically good by altering the hip and abdomen areas. Though the main objective of the girdle is to shape the body by compression, it is also essential that the garment should not create any discomfort or any detrimental effects on the user. Optimum pressure distribution along with the garment with predefined pressure is one of the major issues with compression wear research. Many researchers questioned the reproducibility of the pressure sensor in a human. Hence, to evaluate the pressure distribution among the different parts of the shapewear, the researcher conducted a subjective analysis. They have used three different-sized garments from three different brands. The pressure distribution in the garment was measured at ten different positions namely, front tummy, left front tummy, right-front tummy, left side, right side, left front lower, right front lower, left hips, right hips, and waist level. The results of the study reported that there was a moderate relationship between the tightness feeling and the pressure exerted on the body. The researcher found a significant difference between the tightness rating and pressure feelings on the body parts. The results also suggested that tightness rating was largely influenced by the different other factors like human fat, muscle resilience, and bone structure. As per the participant's feedback, the research optimized the pressure level of all the ten places they have studied. They reported that the participant felt a higher pressure (11.5 mmHg) at the sides, a pressure of 4.5 mmHg at the hip, and 6.5 mmHg at the waist. The overall acceptance of the garment was about 7.5 mmHg for compression shapewear applications [55].

The researcher also developed a statistical model to predict the fabric body pressure and compared the results with the predictions. The results were reported that an overall 77% of correlation noted between the predicted and the actual measurement on humans. The researcher proposed this method as a useful tool for manufacturers to measure the clothing during product development [56].

5.7 CONCLUSION

The main focus of the chapter is to detail the applications of elastane fabrics in the medical field. The effects of elastic yarn properties and elastane fabric's mechanical properties on the end-use requirements were detailed. Specifically, the important properties like the effect of elastane yarn linear density, insertion density, and elastane fabric relaxation on the compression properties were detailed, which is one of the important aspects in the medical field. In this chapter, the potential medical applications like compression bandages, stockings, supports, and shapewear garments are summarized. Various research works performed in the textile and medical area related to the elastane yarn and fabrics were given for better understanding and with practical relevance.

REFERENCES

1. Krimmel, G. (2009). The construction and classification of compression garments. *Template Practise: Compress Hosiery Upper Body Lymphoed*, 2–5.
2. Legner, M. (1999). Medical textiles with specific characteristics produced on flat knitting machines. Medical Text Proceedings International Conference, Bolton, pp. 44–51.
3. Pereira, S., Anand, S.C., Rajendran, S., & Wood, C. (2007). A study of the structure and properties of novel fabrics for knee braces. *Journal of Industrial Textiles*, 36(4), 279–300.
4. Ying Xiong, & Xiaoming Tao. (2018). Compression garments for medical therapy and sports. *Polymers*, 10, 663. doi:10.3390/polym10060663

5. Van der Wegen-Franken, C.P.M., Mulder, P., Tank, B., et al. (2008). Variation in the dynamic stiffness index of different types of medical elastic compression stockings. *Phlebology*, 23(2), 77–84.

6. Rong Liu, Xia Guo, Terence T. Lao & Trevor Little. (2017). A critical review on compression textiles for compression therapy: Textile-based compression interventions for chronic venous insufficiency. *Textile Research Journal*, 87(9), 1121–1141.

7. Oğlakcioğlu, N., Sari, B., Bedez, T., & Marmarali, A. (2016). A novel medical bandage with enhanced clothing comfort. *Materials Science and Engineering*, 141, 120–121.

8. Scott Frothingham. (2019). What to know about compression socks and stockings. https://www.healthline.com/health/compression-socks-benefits (Accessed on November 2022).

9. Understanding Levels of Compression for Stockings. (2020). https://www.legsmart.com/blogs/resources/7032008-understanding-levels-of-compression-for-stockings (Accessed on November 2022).

10. Lim, C.S., & Davies, A.H. (2014). Graduated compression stockings. *Canadian Medical Association journal*, 186(10), E391–E398. https://doi.org/10.1503/cmaj.131281

11. Gokarneshan, N. (2017). Design of compression/pressure garments for diversified medical applications. *Biomedical Journal of Scientific and Technical Research*, 1(3), BJSTR.MS.ID.000309. DOI: 10.26717/BJSTR.2017.01.000309

12. Siddique Hafiz Faisal, Havelka Antonin, Mazari Adnan, & Hussain Tanveer. (2018). Effect of elastane linear density on compression pressure of V-shaped, compression socks. *Industria Textila*, 69(2), 118–127.

13. Burak Sarı, & Nida Oglakcıoglu. (2018). Analysis of the parameters affecting pressure characteristics of medical stockings. *Journal of Industrial Textile*, 47(6), 1083–1096.

14. Wang, L., Felder, M., & Cai, Y.J. (2011). Study of properties of medical compression garment fabric. *Journal of Fiber Bioengineering and Informatics*, 4(1), 15–22.

15. Siddique, H.F. (2019). Assessment of mechanical properties of compression socks using cut-strip method. *Journal of Textile Engineering and Fashion Technology*, 5(5), 228–233. DOI: 10.15406/jteft.2019.05.00206

16. Rong, L., Yi Lin, K., Yi, L., et al. (2005). Objective evaluation of skin pressure distribution of graduated elastic compression stockings. *Dermatology Surgeon*, 31(6), 615–624.

17. Stefanie Reich-Schupke, & Markus Stücker. (2019). Round-knit or flat-knit compression garments for maintenance therapy of lymphedema of the leg? – Review of the literature and technical data. *Journal of German society of dermatology*, 17(8), 775–784, DOI:10.1111/ddg.13895.

18. Homa, M., Marzie, A., Sadeghi, A.H., et al. (2011). On the pressure behavior of tubular weft knitted fabrics constructed from textured polyester yarns. *Journal of Engineered Fibers and Fabrics*, 6(2), 30–39.

19. Rong Liu, Yi-Lin Kwok, Yi Li, & Terence-T Lao. (2010). Fabric mechanical-surface properties of compression hosiery and their effects on skin pressure magnitudes when worn, *Fibres and Textiles in Eastern Europe*, 18(2), 91–97.

20. Mear, J., & Moffatt, C. (2002). Bandaging technique in the treatment of venous ulcers. *Nursing Times Plus (Wound Care)*, 98(44), 3–5.

21. Marie Todd. (2011). Compression bandaging: Types and skills used in practical application. *British Journal of Nursing*, 20(4), 239–241.

22. Finnie, A. (2002). Bandages and bandaging techniques for compression therapy. *British Journal of Community Nursing*, 7(3), 134–142.

23. Alison Hopkins, & Fran Worboys. (2005). *Understanding compression therapy to achieve tolerance. Wounds UK.* 1(3), 29–34. https://www.wounds-uk.com/journals/issue/4/article-details/understanding-compression-therapy-to-achieve-tolerance-1 1

24. Morton, W.E., & Hearle, J.W.S. (2008). *Physical properties of textile fibers* (pp.357–411). London: The Textile Institute.

25. Zhang, X., Li, Y., Yeung, K.W., et al. (2000). Viscoelastic behavior of fibers during woven fabric bagging. *Textile Research Journal*, 70, 751–757.
26. Sengoz, N.G. (2010). Bagging in textiles. *Textile Progress*, 36, 1–64.
27. Kumar, B., Das, A., & Alagirusamy, R. (2014). Effect of material and structure of compression bandage on interface pressure variation over time. *Phlebology*, 29(6), 376–385.
28. Abdelhamid, R.R., Aboalasaad, Brigita Kolčavová Sirková, & Zuhaib Ahmad. (2020) Influence of tensile stress on woven compression bandage structure and porosity. *AUTEX Research Journal*. 20(3), 263–273. DOI: 10.2478/aut-2019–0027
29. Sikka, M.P., Ghosh, S., & Mukhopadhyay, A. (2014). The structural configuration and stretch property relationship of high stretch bandage fabric. *Fibers and Polymers*, 15(8), 1779–1785. doi:10.1007/s12221-014-1779-2
30. Muhammad Maqsood, Yasir Nawab, Jawairia Umar, Muhammad Umair, & Khubab Shaker. (2017). Comparison of compression properties of stretchable knitted fabrics and bi-stretch woven fabrics for compression garments. *The Journal of the Textile Institute*, 108(4), 522–527. DOI: 10.1080/00405000.2016.1172432
31. Abdelhamid, R.R., Aboalasaad, Z.S., Brigita Kolčavová, S., & Amany, A.S.K. (2020). Analysis of factors affecting thermal comfort properties of woven compression bandages. *AUTEX Research Journal*, 20(2). 178–185. DOI: 10.2478/aut-2019–0028
32. Weller, C., Jolley, D., Wolfe, R., Myers, K., & McNeil, J. (2010). Effect of elasticity on subbandage pressure of three layer tubular compression bandages in healthy volunteers: A RCT. *Journal of Wound Care*, 19(10), 417–423.
33. Diana Ališauskienė, & Daiva Mikučionienė. (2012). Investigation on alteration of compression of knitted orthopaedic supports during exploitation. *Materials Science (Medžiagotyra)*, 18(4), 362–366.
34. Cruz, J., Fangueiro, R., Soutinho, F., Fereira, C., & Andrade, P. (2010). Study of compressive behavior of functional knitted fabrics using elastomeric materials. Proceedings of AUTEX World Textile Conference Vilnius, Lithuania, p. 33.
35. Mikucioniene, D., & Milasiute, L. (2017). Influence of knitted orthopaedic support, construction on compression generated by the support. *Journal of Industrial Textile*, 47(4), 551–566.
36. Mikučioniene, D., & Ališauskiene, D. (2014). Prediction of compression of knitted orthopaedic supports by inlay-yarn properties. *Material Science*, 20(3), 311–314.
37. Laima Muralienė, Daiva Mikučionienė, Akvilė Andziukevičiūtė-Jankūnienė, & Virginija Jankauskaitė. (2019). Compression properties of knitted supports with silicone elements for scars treatment and new approach to compression evaluation. IOP Conference Series: Materials Science and Engineering, 500. doi:10.1088/1757–899X/500/1/012016
38. Anand, S.C. (2003). Recent advances in knitting technology and knitted structures for technical textiles applications. In F. Goptepe & O. Goktepe (eds.), Proceedings of ISTEK 2003 Conference, Isparta, Turkey, pp. 96–113.
39. Diana Ališauskienė, Daiva Mikučionienė, & Laima Milašiūtė. (2013). Influence of inlay-yarn properties and insertion density on the compression properties of knitted orthopaedic supports. *Fibres and Textiles in Eastern Europe*, 6(102), 74–78.
40. Compression garments for burns. https://www.healthpartners.com/hospitals/regions/specialties/burn-center/scar-management/ (Accessed on November 17th 2022).
41. Urioste, S.S., Arndt, K.A., Dover, J.S. (1999). Keloids and hypertrophic scars: review and treatment strategies. *Seminars in Cutaneous Medicine and Surgery*, 18(2), 159–171.
42. Macintyre, L., & Baird, M. (2006). Pressure garments for use in the treatment of hypertrophic scars—a review of the problems associated with their use. *Burns*, 32, 10–15.
43. Yildiz, N. (2007). A novel technique to determine pressure in pressure garments for hypertrophic burn scars and comfort properties. *Burns*, 33, 59–64.

44. Li-Tsang, C.W., Zheng, Y.P., Lau, J.C. (2010). A randomized clinical trial to study the effect of silicone gel dressing and pressure therapy on posttraumatic hypertrophic scars. *Journal of Burn Care Residue*, 31, 448–457.

45. Puzey, G. (2002). The use of pressure garments on hypertrophic scars. *Journal of Tissue Viability*, 12(1), 11–15.

46. Jayne, Y., Kim, James J., Willard, Dorothy M.S., Sashwati Roy, Gayle M. Gordillo, Chandan K. Sen, & Heather M. Powell. (2015). Burn scar biomechanics following pressure garment therapy. *Plastic Reconstruction Surgeon*, 136(3), 572–581. doi:10.1097/PRS.0000000000001507.

47. Mustoe, T.A., Cooter, R.D., Gold, M.H. et al. (2002). International clinical recommendations on scar management. *Plastic Reconstruction Surgeon*, 110, 560–571.

48. Atiyeh, B.S., El Khatib, A.M., & Dibo S.A. (2013). Pressure garment therapy of burn scars: Evidence-based efficacy. *Annals of Burns and Fire Disasters*, XXVI(4), 205–212.

49. Jodie Wiseman, Robert S. Ware, Megan Simons, Steven McPhail, Roy Kimble, Anne Dotta, & Zephanie Tyack. (2020). Effectiveness of topical silicone gel and pressure garment therapy for burn scar prevention and management in children: A randomized controlled trial. *Clinical Rehabilitation*, 34(1), 120–131.

50. Joanna Ford. (2017). Rapid review of the efficacy of pressure garment therapy for the treatment of hypertrophic burn scars. http://www.medidex.com/evidence-based-procurement-board-ebpb/865-pressure-garments-in-scars.html (Accessed on August 2020).

51. Goodlad, L.M.E., Kaganovsky, L., & Rushing, R.A. (2013). *Mad men, mad world: Sex, politics, style, and the 1960s*. Durham, NC: Duke University Press.

52. The types of shapewear, https://www.houseoffraser.co.uk/guides/shapewear-guide (Accessed on August 2020).

53. Postpartum Shapewear 101. (2018). The Basics, https://shapermint.com/blogs/news/postpartum-shapewear (Accessed on August 2020).

54. Chen Dongsheng, Liu Hong, Zhang Qiaoling, & Wang Hongge. (2013) Effects of mechanical properties of fabrics on clothing pressure. *Przegląd Elektrotechniczny*, 89(1b) 232–235.

55. Chan, A.P., & Fan, J. (2002). Effect of clothing pressure on the tightness sensation of girdles. *International Journal of Clothing Science and Technology*, 14(2), 100–110.

56. Fan, J., & Chan, A.P. (2005). Prediction of girdle's pressure on human body from the pressure measurement on a dummy. *International Journal of Clothing Science and Technology*, 17(1), 6–12.

6 Elastane in sportswear

6.1 INTRODUCTION

The sportswear market is one of the important sectors in the apparel market, where steady growth has been noted in recent times. The market has been estimated to boom with a compound annual growth rate of 10.4% from 2019 to 2025 holding a market size of 239.78 billion USD in 2018 where the men category accounts for more than 50% followed by the women and kids category [1]. Consumer's knowledge on sportswear recently gained momentum with increased awareness on benefits of fitness activities to get rid of health issues like obesity and mental stress and so the sportswear attracted not only the professional sportspersons but also the normal consumers. The expected benefits of the sportswear fabrics are moisture-handling ability, breathability, easy stretch, and soft feel and ultimately make the wearer comfortable even at extreme workouts and activities [2]. The use of cotton is not preferred and synthetic materials are used due to their lower absorption. This will aid in keeping the garment lighter and non stick to skin even at higher sweating rates during active sports [2]. Moreover, the ability to protect against any impact is also an important consideration in outdoor and performance sportswear which needs the fabrics to have mechanical properties including high strength, impact resistance, abrasion resistance, and tear strength [3]. The type of fibres, structure, and construction parameters influences the thermal and breathability characteristics that are highly desired in the case of sports application [4]. But at the same time, a higher stretch and extreme repeated cyclic pressure application on the clothing is most usual in the case of sportswear. Normally human skins are expected to stretch in some areas like knees and elbows in the range of 35%–45% which can ultimately increase with the intense stretch movements during sports activities. Thus, stretch and recovery of the fabric are the keys to comfort in sportswear [5].

With stretch & recovery, and compression being essential properties, the use of elastane yarn in the fabric to increase its performance in sports became quite common in recent days. Elastane yarns are included in the fabric structures to increase the stretch and recovery properties of the fabric. Elastane in the fabric is also essential to provide compression to muscles for recovery from muscle soreness and to prevent soft tissue injury [5]. The inclusion of elastane can provide stretch and recovery in a controlled manner that can improve the potentiality of the fabrics to be used in sportswear by improving comfort and allowing freedom of body movement [6]. A lower percentage of elastane yarn is used in most innerwear and highly fitted women's wear. Elastane yarns are also commercially used in socks to improve fitness [7]. In the case of sportswear, the percentage of elastane yarn will be more (up to 30%) when compared to casual-fitted garments where the elastane yarn will be in the range of 2%–5% for comfort stretch [5]. Sportswear like gym suits, swimsuits, and

DOI: 10.1201/9780429094804-6

race suits where close-fitting is expected can use elastane fibres with high elongation (stretch 300%–800%) whereas low elongation fibres (stretch 20–100%) suit well for lifestyle sportswear [6]. While considering swimwear, the inclusion level of elastane plays an important role in determining the shape of the body over other parameters like fabric structure/weave [5]. Moreover, better mechanical properties like tenacity and modulus were also noticed with elastane fibres [6].

Elastane has the versatility of providing elastic properties when made into different fabric types, including wovens, non-wovens, circular, and flat knits. However, elastomeric woven fabrics are not popular because of their high density and low porosity compared to knits. Elastomeric knitted fabrics are most commonly used in sportswear [6]. The need for elastane fabrics in sportswear got accelerated when the performance of athletes was noted to be improved while using compression garments. The proper selection of compression levels for a particular sport can improve force, power production, and proprioception [6]. The method of manufacturing also plays an important role in the performance of elastane fabrics meant for sportswear. Researchers have reported that the use of spandex-plated cotton fabrics to have better applicability for athletic sportswear than the spandex core cotton spun fabrics However, the use of elastane does have certain issues. The increase in the elastane content can affect thermo-physiological characteristics as the elastane can make the structure tighter and potentially affect the air permeability which is essential in the case of sportswear [7].

This chapter consolidates the various requirements of sportswear fabrics and the use of elastane yarn in sportswear. The first part of the chapter details the material requirements of active sportswear in the aspect of stretch and recovery. An analysis of the various stretch levels and comfort properties required for a sportsperson was detailed. The second part of the chapter outlines the necessity of elastane fibre in sports textiles. It particulars the elastane-incorporated compression garments in sportswear with their benefits and limitations. The last part of the chapter elucidates the impact of elastane on the stretch, recovery, and comfort aspects of sportswear fabric.

6.2   ACTIVE SPORTSWEAR

The term active sportswear covers a wide range of apparel in the industry from professional sports apparel to casual wearers. In the case of professional sports, the main requirement of active sportswear is human comfort. In recent years, the significant invention of innovative materials and design has increased the performance of the wearer by controlling the thermo-physiological comfort of the fabric. The main requirements of sportswear are provided in Figure 6.1. In general, there are three types of sports players:

    i. Professional sports players, who mainly focus on victories and record-breaking. These professional players mainly require functional power with aesthetic appeal.

    ii. Future professionals are the amateur players who are actively involved in the sports clubs, schools, and local sports, who have a higher potential in the relevant sports and going to be future professional players. They require a minimal functional value at an affordable cost.

iii. Common consumers are the people who use sports activities for the benefit of health, hobbies, exercise, and social contacts. These categories require a minimal physical function with higher athletic, comfort, and sensitivity along with easy care.

Due to the increased awareness of health and physical fitness among the public, sportswear become one of the everyday clothing during morning walks, jogging, yoga, stretching exercises, and daily fitness activities because of the quality comfort of sports clothing. A research report mentioned that only 30% of the sportswear manufactured is utilized by active sportswear persons [8]. As sportswear is worn next to the skin it plays a vital role in the athlete's physiological comfort and so on the performance of the athletes. The important attributes of the functional comfort of sportswear are provided in Figure 6.2 [9].

In sportswear, most of the time, the comfort properties of the garments were not given any importance. However, researchers identified that the discomfort caused by clothing negatively affects the performance of the wearer. So, researchers reported that identifying and optimizing clothing comfort with respect to sports will enhance the performance of the individuals. Particularly in warm and hot conditions, higher body sweat causes most of the discomforts and water loss to the body. Concerning the hot or warm conditions, the important features are (i) minimizing any negative effects on heat loss and (ii) maximizing any positive effects on heat loss.

The negative effects of the clothing systems are resistant to heat and water vapour loss during athletic activity. This can be effectively reduced by selecting appropriate material with a low resistance to moisture, air, and heat. In this case, the design aspects of the clothing also play a vital role. The proper selection and

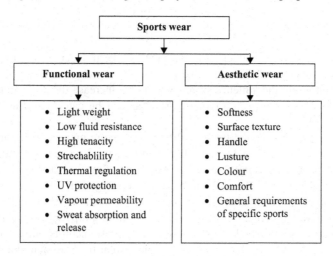

FIGURE 6.1 The requirements of sportswear.

Functional Comfort of Sports Wear

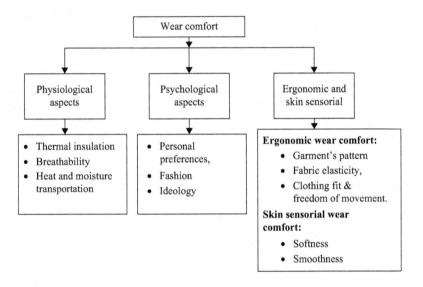

Rapid Moisture Absorption Conveyance Capacity	Easy Care Performance	Dimensionally Stable even when wet
Absence of Dampness	Rapid Drying to prevent Catching Cold	Light Weight
Optimum heat and Moisture Regulation	Low Sweat Absorption	Good air & Water Vapour Permeability

FIGURE 6.2   Attributes related to the functional clothing comfort of active sportswear.

Wear comfort

Physiological aspects

Psychological aspects

Ergonomic and skin sensorial

- Thermal insulation
- Breathability
- Heat and moisture transportation

- Personal preferences,
- Fashion
- Ideology

Ergonomic wear comfort:
- Garment's pattern
- Fabric elasticity,
- Clothing fit & freedom of movement.

Skin sensorial wear comfort:
- Softness
- Smoothness

FIGURE 6.3   Different types of wear comfort and their influencing parameters.

effective design of the product will also increase the transfer of water vapour and heat through the material. The effective selection of both the design and material will help in maintaining the athlete's skin dry and also increase the performance of the wearer. The comfort aspects of the sportswear can be categorized as follows in Figure 6.3 [10].

In sports applications, the breathability and moisture management ability of the fabric is very crucial. The breathability mentions the heat and air transport and moisture management which determines the ability of moisture transmission in both the liquid and vapour form. All these parameters together influence the thermo-physiological comfort and alter the microclimate between the fabric and the skin. At the extreme sweating conditions, the fabric absorption property alone will not help the wearer to feel comfortable, it is the liquid transmission ability of the fabric which makes the wearer feeling dry and comfortable. Sweat transmission and absorption avoid the discomfort of the fabric sticking to the skin. Another important aspect of the fabric that has a significant influence on the sensorial comfort of the clothing is the surface parameters such as friction, roughness, and softness. In sports applications, in extreme wet situations, poor selection of fabric may cause chafing to the wearer and it is one of the common issues in sportswear. Another kind of issue that is very often faced by many athletes and sportsperson is ergonomic discomfort. It is defined as the fit and free movement of the apparel which is worn very next to the skin. In wet conditions, the fabric's behaviours differ from the dry condition and cause ergonomic discomfort by restricting the free movements of the body during sports activity. Hence, in these applications, elastic fabrics are commonly used as it possesses higher stretch and recovery, improved fit, desired shape at extreme body movements, and adequate room for ease.

6.3 FIBRES USED IN SPORTSWEAR

In the post-industrial revolution, the use of cotton was reduced significantly in garment production due to the advent of synthetic fibres like polyester, nylon, polypropylene, etc. The micro denier polyester fabric was widely adapted in sportswear applications due to its lightweight, cheap to produce, dye fastness, durability, easy-care properties, quick-drying, hydrophobic in nature, and wicking ability. Micro denier polyester is often blended with other natural fibres mainly to extract its benefits to maintain moisture management and durability. The stretch and recovery properties are essential properties for sportswear. Hence, elastane fibre occupied a significant space in sportswear applications. The elastic nature of filaments is used in sportswear to compress muscles, offer stretch for body movements, and support in recovering from muscle soreness. A wide range of sportswear products such as foundation garments, swimwear, base layer products, and compression tights are widely made of elastane. Table 6.1 represents a few commonly used fibres in sportswear and their special properties along with their application area [11-15].

Elastic or stretchable fabric can be produced using available knitting and weaving technologies. The elastic yarns will be most of the time wrapped in the core or braided in the core with the help of different methods as discussed in Chapter 2. However, the elastic yarns can provide the inherent stretch and recovery effect for widely varying items regardless of whether it is a bare or a combination yarn. Based on the extension percentage, the yarns can be classified into low (20%–150%), medium (150%–390%), and higher elastic fibres (400%–800%). The fibres like nylon 6, nylon 66, and cotton are also used for the manufacturing of elastic yarn by using different techniques as mentioned [16] in Figure 6.4.

Apart from extensibility and elastic recovery, elastic hysteresis is also an important phenomenon for elastic fabrics. Hysteresis reflects the stress relaxation of elastic fabric when it has been subjected to repeated stretching and recovery. It is also important to consider the impact of elastane yarn in a normal fabric in terms of physical properties other than its functional requirements. The following sections will detail the effect of the inclusion of elastane on the physical properties of woven and knitted fabrics.

TABLE 6.1
Fibre types and their special properties of sportswear

s. No.	Fibre name	Special properties	Application
1	Cotton	• Softness and comfort • Absorb and retain moisture • When wet, cotton garments cling to the skin causing discomfort	• Not recommended for skin contact wear but used as a second layer to absorb sweat and helps to evaporating of the absorbed sweat,
2	Viscose rayon	• Higher (13%) moisture regain • Higher moisture absorption than cotton	• Not recommended for skin contact wear • Can be used as the outer layer of knitted hydrophilic portion of the twin layer sportswear
3	Nylon	• Lightweight, high strength, and softness with good durability • Higher moisture regain than polyester • Better wicking behaviour than polyester	• Most often used in tightly woven outerwear • Breathable knitted fabrics
4	Micro-polyester	• Dimensional stability • Excellent resistance to dirt, alkalies, decay, mould, and most common organic solvents • Heat resistance or thermal stability • Low moisture absorption • Easy-care properties	• Base fabric in active wear • Chemical treatment necessary to wick moisture as a base later
5	Polypropylene	• Polypropylene can wick liquid moisture • Moisture vapour can still be forced through polypropylene fabric by body heat • Polypropylene provides insulation • Higher moisture management and good thermal characteristics	Keeps the wearer warm in cold weather and cold in warm weather
		Regenerated fibres	
6	Tencel	• The moisture management capability of tencel is unique • Provides excellent performances in sports • Excellent moisture absorption	Skin contact layer
7	Bamboo	• Excellent wet permeability • Moisture vapour transmission property • Soft hand, better drape, easy dyeing • Inherent anti-bacterial property • Inherent ultraviolet protection	• Underwear, tight t-shirts, and socks • Summer clothing

s. No.	Fibre name	Special properties	Application
8	Soybean	• Superior soft hand • Higher moisture absorbency and permeability	Innerwear
		Technical Fibres	
9	Killat N	• It is a nylon hollow filament • The hollow portion is about 33% of the cross-section of each filament. • Good water absorbency and warmth retentive property	Innerwear Sportswear Athletic wear
10	Lycra	• Lycra, a synthetic fibre composed of 85% segmented polyurethane • Stretch and recovery	Swimwear, active sportswear, floor gymnastics
11	Hygra	• A sheath-core type filament yarn composed of fibre made from water-absorbing polymer and nylon • Absorbs 35 times its own weight of water. • Quick releasing properties • Superior antistatic properties even under low wet conditions	Athletic wear, skiwear, golf wear, etc.
12	Dacron	• Four-channel polyester • Engineered to move moisture and speed the evaporation of perspiration • Superior fabric for wicking action, drying time, moisture absorption, and transport	Athletic wear

6.4 PROPERTIES OF ELASTANE KNITTED AND WOVEN FABRIC

Many research works are performed on the physical properties of elastane knitted fabric. The main concept in which the elastane yarn is introduced in the knit fabric is to increase the fit, higher level of stretch, and more dimensional recovery. The process of combining elastane with a natural or manmade yarn in the knitting process is called plating. Plating means the simultaneous formation of one loop from two yarns so that one yarn will lie on the face of the fabric. The details of production methods and properties of the elastane fabric were discussed in Chapters 2 and 3. In application point, the elastane-plated knitted fabrics are mainly used in sportswear, underwear, swimwear, and tights. Additionally, these fabrics also find their application in the field of the medical aspects as corsets, compression bandages, and support stockings in sports as detailed in Chapter 5. Due to the changes in trend and increase in consumer knowledge on clothing choices for different occasions, the usage of elastane-plated fabrics was also increased tremendously in recent years. The

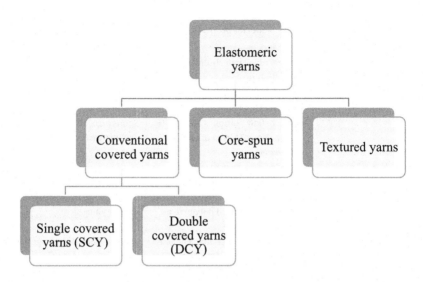

FIGURE 6.4 Classification of elastomeric yarns used in textile application.

industry anticipates an annual increase of 8%–10% in the elastane fabric consumption among the consumers for the coming years. The elastane knit fabrics may have elastane in every course or alternating courses known as full plating and half plating respectively, usage of the yarn is based on the application requirements. This section details the effect of elastane content on the comfort aspects of the textiles while in the application.

Knitted fabric produced with elastane yarn has the largest tension values under a constant draw ratio which gives the highest weight, wales/cm, courses/cm, stitches/cm, thickness, and lowest air permeability values [17, 18]. The weight and thickness of full plating cotton/spandex fabrics are higher, but air permeability, pilling grades, and the degree of spirality are lower than half plating cotton/spandex and cotton fabric [19, 20]. In the case of single jersey fabric, the full plating technique offers a very uniform appearance and the half-plated fabrics create ridged effects [21]. The increment in elastane increases the crease recovery of the fabric and also the fabric weight [22, 23]. When the low-stress mechanical properties of cotton/spandex and polyester/spandex blend knit fabrics were analysed, a higher elastic recovery was reported due to higher elastane content [24]. The dry, wet, and full relaxation conditions significantly altered the stitch density of cotton/spandex interlock structures than 100% cotton fabric [25]. Similar results were also noted with 1×1 rib under dry, wet, and full relaxation conditions [26]. An increase in elastane percentage increases the widthwise recovery of fabric. A higher recovery was noted with full-plated fabrics. The air permeability value is indirectly proportional to the elastane content and a positive correlation was reported with bursting strength [27]. Out of all the parameters, dimensional characteristics and related properties deteriorate more than the few positive aspects when elastane is introduced into the fabric [28, 29].

The woven fabrics with elastane yarns are one of the recent developments in the functional apparel sector. The elastane included fabrics possess higher extensibility, elasticity, a high degree of recovery, and good dimensional stability. These fabrics are mainly used in the areas where comfort and fit are the main focus. The limited stretch and recovery properties of the conventional woven fabric are the main reason for the development of elastane blended woven fabrics. The addition of elastane into the fabric directly increases the functional aspects of the fabric like the stretch and dimensional recovery, comfort, and wrinkle resistance. One of the important issues with the woven fabric is the poor stretchability and recovery nature at cycling loading. Due to the cyclic loading, the fabric's mechanical properties will reduce. This is generally known as fatigue. The use of core-spun yarn containing lycra has become one of the effective ways to impart suitable stretch and recovery properties to fabrics without undesirable deformation of the garment during its service life. The research identified that the introduction of elastane yarn in the weft direction reduces the tensile and tearing strength values of denim fabrics when compared to the weft yarn without elastane content. Further, the increment in the elastane content also reduces the bagging effect of the woven fabric and increases the elastic recovery and stiffness of the fabric [30]. The addition of elastane yarn in the fabric had increased the contraction, crease recovery, and flexural rigidity with increasing elastane drawing ratio. However, it was reported that the rest of the fabric properties like air permeability, tearing strength, and other physical properties were significantly reduced [31, 32]. Finer core-spun lycra yarn improved the smoothness and total hand values of the elastane fabric, higher stretch, and recovery than the coarser yarn [33, 34]. An increment in elastane ratio has increased the fabric extensibility and rate of air permeability [35]. Due to the increased breaking extension, elastane fabric reduced its deformation characteristics [36–38]. The elastane content in the core-spun yarn has a strong positive correlation with fabric tear strength and recovery after stretch, and a strong negative correlation with fabric tensile strength [39, 40].

6.5 SPORTSWEAR REQUIREMENTS

Stretch and recovery of the fabric are one of the important requirements in the case of sports applications. A simple and normal movement of the body like sitting, bending the elbow or knee stretches the human skin in those places around 50% (Figure 6.5). This will help us to understand the level of stretch required for active sports. When the sports person's outfit is not extensible like skin, it results in movement restriction for the wearer, and also in the subsequent usage the garment shape changes and results in a reduced performance than the expected level. To maintain the level of elasticity for different applications, elastane is introduced in fabric production. The introduction of elasticity to sports apparel not only increases the performance of the athletes but also reduces the chances of getting an injury, muscle fatigue, and friction drags.

Irrespective of the structure of fabrics, elastane can be used with all types of fabrics. Elastane yarn will increase the stretch and recovery properties that will ultimately increase the comfort of the garments by providing unrestricted movements and strong shape retention characteristics. Hence, the application requirements decide the quantity and quality of the elasticity required. In general, elastane is used

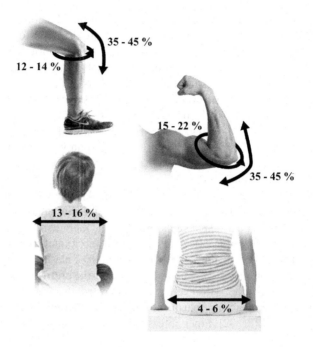

FIGURE 6.5   Skin stretch level at different parts of the body.

from 2% to 30% commercially based on the application requirements. Human wear comfort of the apparels can be classified as follows [41]:

1. **Thermo-physiological wear comfort**: It is related to the heat and moisture transportation through the fabric and clothing material. It is usually measured in terms of thermal conductivity, breathability, and moisture management characteristics.
2. **sensorial wear comfort**: It is the mechanical sensation between the skin and clothing. It may be smooth and soft known as pleasant perception or unpleasant like scratch, stiff, and clings to sweat.
3. **Ergonomic wear comfort**: It is mainly based on fit and ease. The garment pattern and elastic nature of the fabric are the responsible factors for ergonomic wear comfort.
4. **psychological wear comfort**: It is affected by fashion, personal preferences, ideology, etc.

Technically, elastane yarn can improve the ease of the cloth and enhance the ergonomic comfort of the fabric. But at the same time, the major problem with elastane yarn is that they are synthetic and hydrophobic in nature. As a result, they do not absorb moisture and practically, they are not wettable by any liquid. Hence, the inclusion of elastane yarns in a fabric significantly reduces the thermophysiological wear comfort of the fabric. Still, there are perceptions among the manufacturers and fabric producers that the inclusion of elastane and microfibre into the fabric alone will make good sportswear. But the comfort aspects of any apparel are not defined by anyone parameter like elastane. The comfort properties required for every single sport differ significantly and all the physiologically relevant construction parameters can be engineered carefully for specific applications to enhance comfort.

6.6 REQUIREMENTS OF HIGH-ACTIVE SPORTSWEAR

In high-active sport, the requirements of sportswear depend upon the nature of the sports, physical activity involved, and climatic conditions. In general, sports like tennis, soccer, running, and jumping are classified as high-active sports as they are played for a shorter time with maximum physical activity. During the high active sports activity, the human body generates a high level of metabolic heat, and hence the body core temperature increases by 1.5–2°C. To control this generated heat, as a protective mechanism, the body sweats and produces a cooling effect through the skin [42]. Bartels reported that during the high-active sports, the body generates sweat at very higher level up to 2.5 L/h [43]. Hence, the main requirement of the high active sportswear is to absorb the sweat faster along with rapid drying ability. In addition, as these sports require extreme body movements, the apparel should also have higher elastic recovery and enough freedom for movement with a good fit. In some cases, sportswear is also used as a compression fabric to increase the performance of an athlete [10].

6.7 ROLE OF ELASTANE IN SPORTSWEAR

As mentioned in the previous section, the main purpose of using elastane in sportswear is to impart elasticity and elastic recovery. Generally, sportswear can be categorized into four groups as provided in Figure 6.6.

As per Figure 6.6, performance sportswear is highly technical-oriented clothing that enhances sport performance with special functionality. It is produced in the lowest volume and highest price range, whereas basic sportswear is cheaper and more stylish while retaining as many of the material attributes as possible. Sports leisurewear is a replica of performance sportswear, worn at home, and is sold in higher volume at a much smaller price. Sports-fashion clothing, as the terminology mentions, is the trend, which adopts the latest styles from sports performance and basic wear to fashion wear [44].

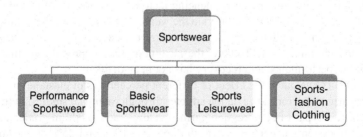

FIGURE 6.6 Classification of sportswear.

Performance sportswear are garments that are used in highly active sports like tennis, soccer, and athletics. These sports are highly active and during the play, heat stress is a great concern as the sportspersons develop a higher level of metabolic heat ranging from 800 to 1300 W [10]. This increase in body heat directly generates body sweat proportionally to maintain a balance in core body temperature. Research reports mentioned that the sweating of the body passes the body heat to the environment through the heat of the vaporization of water and subsequently cool the body. It is also reported that the sweat generation can be as high to achieve this effect [44]. One of the main requirements of high active sportswear is a high stretch and elastic recovery to provide sufficient fit and freedom of movement to the wearer. Also, in many athletic activities like running and weight lifting, the elastic fabric is used to create compression for the athletes concerning anatomical fit and muscle fatigue. The following sections detail the various benefits that elastane tenders to the sportswear application.

6.7.1 COMPRESSION PROPERTIES

In performance sportswear, compression garments (CG) play a vital role. The use of compression garments in sportswear is increased due to the possible benefits associated with it. During the physical activity, the garments apply pressure on particular areas of the body to reduce the discomfort induced by the activity and so to improve the wearers' performance. The research reports mentioned improvements in venous [45] and arterial blood flow [46] which could contribute to the wearer's performance and recovery due to compression garment usage. The main materials used for the manufacture of compression garments are polyamide, polypropylene, and nylon, with elastane. The compression effect on the compression garment is mainly based on the elastane fibre used in it. The percentage of elastane and quality defines the compression parameters, such as elasticity, robustness, tact, and thermo-physiological balance, among others. The garments have been manufactured in different shapes for various parts of the body like the foot, knee, hip, thigh, and upper limbs, or larger body areas, for example, compression T-shirts or compression trousers. Figure 6.7 represents the various factors that influence the comfort characteristics of compression garments.

The compression garments are mainly produced from the synthetic fibre with elastane blend [47]. However, the proportion of elastane in its product is based on the application and requirement of the particular sports. The type of finishing used on the material is the second major influencing factor. The majority of the compression garment manufacturer claims to provide the wearer with enhanced blood flow, better muscle oxygenation, reduced fatigue, faster recovery, reduced muscle oscillation, and reduced muscle injury [48]. To date, the ideal compression pressure required to be beneficial to performance and recovery has not been defined.

Researchers evaluated the effect of different percentages of elastane in the compression garment performance on the athletes through wear trial studies. In the study, three different types of compression garments, low-grade, medium grade, and high-grade compression garments were used with 12%, 18%, and 25% of elastane content. The compression garments were used with athletes and their uphill running

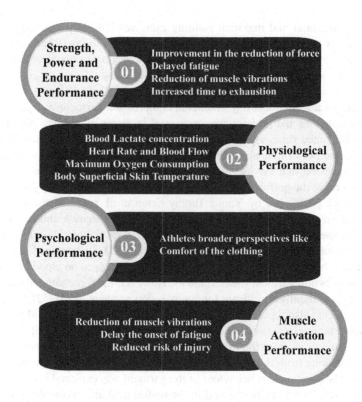

FIGURE 6.7 Factors influencing the performance of the compression garments.

performance was measured. The results were measured in terms of athletes' mean performance and rate of perceived exertion during the experiment. The findings revealed that there is no statistical improvement observed in the running performance of the participants between the different compression garments. The lowest median and average performance times were found in the medium-grade compression garment. The median time performance for the medium-grade compression garment was 1.9% and 1.1% better than in the low-grade CG and the high-grade reverse-compression garment. A similar result was noticed in the case of perceived exertion during experimental running between the compression garments. There are no significant differences observed between the garments. The results also mentioned that increasing pain tolerance was observed in the group of athletes with medium grade compression garments than the other two [49].

Other researchers evaluated the performance of runners with and without gradient compression bandage in the calf area. The researcher used 85% Polyamide and 15%Elastane mixed compression bandages. The experiment was performed among 21 men runner to run on the treadmill with and without the below-knee compressive stockings with ten days of recovery time. The results reported that due to the application of compression garments in the calf area significantly improved the running performance of calf muscle at different metabolic stages during an all-out task. They had concluded that the compression stockings were effective for enhancing performance during sub-maximal and maximal running exercises [50]. Few researchers evaluated the effect of a lower-body compression garment on the performance after the 40-kilometre cycling trail. The researchers used a graduated compression garment made of 76% Meryl elastane and 24% lycra with a non-compressive garment made of 92% polyester and 8% spandex. The athletes were allowed to perform a 40 km cycling and on a subsequent day graduated compression garment was provided and the performance of the athletes was measured in terms of average power output, mean performance time, and mean oxygen. A similar experiment was again conducted and the performance was measured with a non-compressive garment. Their findings suggested that the use of compression garments after 40 km cycling was likely to increase the performance of the athlete in the subsequent day [51]. A similar research work performed by Yunus Turgay Erten et al. reported the use of compression garments in the reduction of athletes' muscle soreness and recovery. The researchers used a compression garment made of 64% polyamide, 17% polyester, and 19% elastane with a compression value range of 20–30 mmHg. Their findings reported that there is no improvement or beneficial effect seen on blood lactate levels within compression garment measurements. However, the force generation capacity at recovery is statically significant for compression garment usage [52]. Some of the other researchers also reported that the use of compression garments did not improve sports performance or significant effect on sports performance [53–56].

At the same time, from the literature, it also can be noted that such performance analysis was done from the textile science point of view. The effect of elastane percentage on the compression behaviour of the garment was explored in real-time situations. The research work performed in the textile field and sports field still stands apart. In the recent research work, the effect of elastane yarn linear density on the compression performance of the socks was analysed. The researcher evaluated the effect of elastane linear density of different yarns namely core-spun yarn, double covered nylon filament yarn, and plaiting yarn. Their finding suggested that the increment in elastane linear density significantly increased the compression pressure of the socks for core-spun yarn, double covered nylon filament yarn. In the case of plated yarn, there is no significant difference noted in their study [57]. The role of other fibres blended with elastane yarn in the compression garment was analysed by a few researchers. They have analysed the effect of other fibres like cotton and nylon on the compression ability and comfortability of the compression garment. Their research findings revealed that the content of nylon and cotton did not have any influence on the pressure exerted. However, comfort characteristics depend on the percentage of the cotton and nylon used [58].

Barhoumi et al. evaluated the effect of manufacturing parameters of knitted compression fabric such as knitted structure, elastane percentage, elastane yarn count, and stitch length on the interface pressure created by the garment. They analysed the effect of these parameters on the pressure using a FlexiForce® force-pressure sensor. Their findings were reported that the elastane percentage is one of the crucial parameters which has a major impact on the pressure applied. They mentioned that the second most important parameter is the count of the elastane yarn. The increment in the count increased the pressure of the garment due to the tight structure. The plain fabric was capable of producing more compression pressure than the pique structure.

With respect to the stitch length higher the stitch length lowers the pressure generated [59]. The expected performance benefits of the elastane compression garments are provided in Figure 6.8.

Studies analysed the possibilities of predicting compression performance through yarn properties with the help of six groups of double-covered elastic threads. Their findings revealed that the increment in the elongation has a significant influence on the tensile force imparted by the garment. Similarly, the linear density of elastic core yarn was also observed significantly. They had concluded that when the influence of inlay-yarn linear density and number of threads on tensile force is determined, it is possible to predict the influence of inlay-yarn properties on compression properties [60]. Recently a researcher developed localized compression clothing using polyester, lyocell, and elastane yarn and evaluated the muscle movements through ECG. The results of the study reported that an increase in elastane percentage improved the thermal resistance due to the compactness. This also led to a reduction in air and moisture permeability results. The increment in elastane percentage positively influenced the spreading rate and drying rate and negatively influenced the wicking. The developed products effectively control the muscle movement up to 90% and increase the athlete's performance to the level of 9% [61].

While analysing the effect of elastane linear density on fabric stretch and recovery, it was found that elastane yarn count impacts fabric recovery dominantly. Researchers reported that woven fabric with higher elastane linear density and higher pick density will produce more pressure in a compression garment. The authors also developed a model to predict the fabric contraction, garment pressure, stretch, and recovery properties of compression garments [62]. Compression stockings with different compression levels (low, medium, and high) were analysed and reported by Ali et al. [63]. They found that the use of compression garments significantly improves

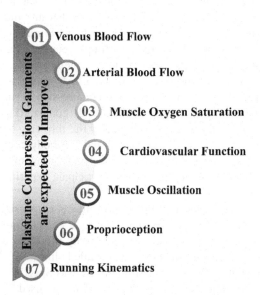

FIGURE 6.8 Expected benefits of elastane compression garments.

the jump heights before and after exercises with the lower and medium tight stockings than the control samples. A similar positive result by other researchers showed a slight improvement in the performance and a higher level of reduction in muscle soreness. The research also supported the elastane imparted compression garment assistance in the posture support and post-training muscle recovery [64]. An increment in the blood flow on the lower limb after the use of compression garments was also reported by Brophy Williams et al. [65]. In a comparative study, the researcher noted that the effectiveness of the compression socks in reducing the leg swelling of the participants. A performance improvement in the compression socks was reported by other researchers on the baseball and golf games. Division I collegiate baseball pitchers' and golfer's performance was evaluated for their accuracy and performance in their respective sports. A nylon/elastane blended torso apparel used in this analysis showed a significant improvement in the sports people's performance. A significant improvement in the pitching accuracy of the baseball pitchers was noted; however, no differences were noted in the ball velocity. In the case of golf, though no differences were noted in the driving distances, a significant improvement was noted in the driving accuracy. In the perception, baseball pitchers felt a greater level of enjoyment in wearing the garment. Whereas, golfers reported a significant improvement in comfort levels [66].

Tomas Venckunas et al. measured the hemodynamics and running performance of the lower body compression garment with 74% polyamide and 26% elastane. The pressure developed by the garment was noted as 17–18 mmHg in the calf and upper thigh region, whereas the control garment provided a 4 mmHg on those portions while standing. Through a training session, the researcher measured the heart rate, leg blood flow, blood pressure, calf muscle tissue oxygenation, and skin temperature after running for 30 minutes. The results showed an increase in skin and core temperature under compression garments significantly due to the lower air circulation under a compression garment. The heart rate of the athlete also did not change during the session after the usage of the compression garment. Similarly, no cardiovascular or metabolic effects were noted. Similarly, no improvement in the running economy also noted with the usage of the compression garment. Outset, the researcher did not report any significant improvement other than body temperature increment with the participants [67]. The effect of full-body compression garments was analysed on the sports performance of the wearer. The study compared the different brands and measured the accuracy and performance in the throwing test. The results did not show any difference between the control and full-body compression garments [68]. The compression socks, tights, and whole-body garments also did not show any performance benefits among the fifteen participant experiments performed by Sperlich et al. [69].

However, few practical and real-life improvements emphasize the role of elastane compression garments in sports. The skin-tight compression bodysuit used in the 2000 Olympic Games in Sydney were primarily produced to reduce drag and increase full-body movement. The post data revealed almost 85% of the gold medal winners wore that suite [70]. Another similar case was Speedo's LZR Racer, a low drag ultra-thin water repellent fast-drying suit that was used in the 2008 Beijing games. The top medal winners including US swimmer Michael Phelps, who bagged

8 gold medals wore the suits [70]. The use of these suits made the athletes break their records and that made a ban on the suit in the 2010 Olympic Games [71].

An analysis revealed that among evaluated studies, the use of compression garments significantly reduced the delayed onset muscle soreness (DOMS) and moderately enhanced the recovery of muscle strength and power post-exercise [72]. There are several reasons said, the improvement in the sports performed confirmed by few other research studies also in the subsequent analysis [73]. One of the most recent metadata analyses published also reported that the performance variable like time to exhaustion and running economy were highly influenced by the use of compression garments. Though they are not the direct representation of sports performance like success and failure, it influences or supports the performance significantly [74]. The major factors that influence the sport performance of the elastane compression garment are provided in Figure 6.9.

6.7.2 STRETCH AND RECOVERY

Next to the compression properties, the most expected property of the elastane fabric in sportswear application is stretch and recovering ability. When the above-mentioned properties are correctly selected, the expected performance of sports compression fabric is mainly on the stretch and recovering ability of the fabric at sports activity. In performance sports, the active wear fabrics must have the property called dynamic elastic recovery, which helps in analysing the garment's response to instant body movements. The elastane content in the fabric actively retracts back after every stretch and restricts the bagging behaviour of the textiles. These two properties were interconnected and mainly influenced by the fibre type, elastane content in the fabric,

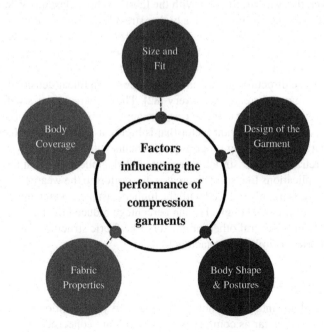

FIGURE 6.9 Factors influencing the sport performance of the compression garment.

and the fabric type. Previous research results noted that all the fabric properties and method of elastane incorporations in the fabric had their impact on the performance. As far as the stretch properties of the sportswear are considered, an increase in elastane percentage in the fabric significantly increases the stretch and recovery properties irrespective of the fabric type [75–77]. Bi-stretch woven fabrics showed the better stretch and recovery properties than a weft-alone elastane inserted fabric. When the knitted fabrics are considered, a higher stretch is noted with the full-plated knitted fabric than the half-plated knit structure [77]. Though it was reported that increment in the elastane percentage increases the stretchability, the research reports also mentioned that increment in the elastane yarn linear density reduces the stretch and recovery ability of the fabric [78]. The reduction in the stretch is mainly attributed to the filament bulkiness while increasing the count. Similarly, it was also reported that the increment in the elastane draw ratio and pick density (picks per inch) of weft elastane yarn reduces the stretch and recovery properties of the fabric [79]. These findings were elaborately discussed in the previous chapter (3), and it reveals that an increase in elastane percentage is mainly necessary to provide necessary stretch properties to active sportswear.

Technically, the dynamic work recovery refers to the power gain of the wearer after a maximum stretch. An increment in the elastane percentage in the fabric significantly increases the dynamic elastic and work recovery characteristics of the sports textiles. The elastane-incorporated fabrics showed a higher extension and lower stress value on both wales and coursewise direction than that of 100% cotton fabric [80]. A 20% higher dynamic work recovery in the wales direction and 15% increment in the course direction were noted with the elastane fabric during the stress application. However, this was not the case with the fabric without elastane. When the stress applied on the wearer was measured, a higher stress level was noted in the full-plated fabric than the core-spun elastane knitted fabric [81]. Research on elastic plated cotton knitted fabric revealed that in the wales direction, an increase in spandex yarn linear density does not seem to affect growth and there is no real trend for stretch. In the coursewise direction, an increase in spandex yarn linear density decreases the stretch and increases the elastic recovery [82]. These characteristics will deliver the high-level stretch during the extreme pressure application in active sports like tennis, running, etc. Another important factor that is highly associated with the stretch and recovery character is bagging properties. Garments with poor stretch-ability exhibit a very poor retraction after stretch by repeated load application. That too in the case of extreme applications like sports it creates discomfort to the wearer. Again the bagging property was mainly reduced with the higher elastane content and elastane draw ratio in the sportswear. Though elastane percentage reduced the bagging property, a significant effect of several other parameters like fabric structure, fabric anisotropy was also reported as influencing factors [83].

6.7.3 Comfort Characteristics

Though several advantages are noted with elastane yarn addition in the woven and knitted structures, as far as comfort characteristics are concerned it provides only a negative impact. Due to the hydrophobic structure of elastane yarn, several comfort characteristics negatively improve and reduce the wear comfort. Hence, several researchers advised having a minimum percentage of elastic yarn in fabric. The addition of elastane primarily increases fabric thickness and compactness along with weight. An increase in the elastane percentage (either by density, draw ratio or linear density increment) significantly reduces the air permeability and water vapour permeability of the textile [84]. In the case of wicking properties, a contradiction was noted among the researchers. While few reported a significant reduction due to structural tightness and others reported an increment (Chapter 3). The reduction in the wicking and permeability values significantly affects the wearer's comfort by restricting the sweat transfer from the skin to the atmosphere. In the case of moisture management analysis, a similar reduction in the moisture characteristics was reported. The importance of different elastane proportions, elastane stretch, and twist multiplier of Core-spun elastane-cotton blended yarn in knitted fabric manufacturing was reported by other researchers. They have statistically evaluated the influence of the various fabric parameters on comfort properties. They reported that, with the increase in elastane percentage and elastane stretch, the fabric becomes compact and thick with higher thermal resistance and reduced permeability to air and moisture vapour. The increment in elastane percentage has a negative impact on the wicking and OMMC of the developed fabric [85]. An identical result was reported by Amany Khalil et al. They compared the moisture characteristics of elastane plated, core-spun, dual core-spun yarn and reported that the overall moisture handling ability of the fabric reduces when the elastane was included. Out of all the selected fabrics, a higher OMMC value was noted for 100% cotton followed by full-plated knit, core-spun knit, and dual-core knit fabrics [84]. As the moisture management indices were noted as a significant parameter that influences wearer comfort, the reduction in OMMC value depicts the discomfort. Thus, a higher elastane percentage shows lower absorption, water spreading, and transferring ability of the skin moisture and makes the sportspersons feel damp and wet. Thus, a higher elastane content increases the discomfort. Hence, it is important to have an understanding of the elastane for stretch requirements and also with comfort properties.

Positively, few researchers reported an increment in the thermal conductivity that provides a better help in transferring body heat successfully. Similarly, the surface and handle properties of elastane knitted fabrics analysed, a significant reduction in total hand value (THV) of the elastane fabric was observed [48]. An overall, when comfort properties were concerned, only the thermal conductivity of the elastane fabric was noted to have a positive effect. Whereas, other properties like water vapour permeability, air permeability, wicking, OMMC, and THV of the fabric get reduced with an increase in the elastane content. Hence, it is very imperative to understand the real requirement of the particular application and design the fabric accordingly. One design or one fabric type may not solve all requirements of the sportsperson and hence a tailored approach is recommended to have balanced properties including comfort and stretch and recovery. A sport that requires extreme stretch and recovery like splinting and running may be required to use the garment for a shorter time compared to a sport like football and rugby. Hence based on the amount of physical activeness or by considering the duration of the sport that performed a design can be improved. Though several types of research were performed in the compression garment on sports application as reported in the previous section, no research was found to relate these textile characteristics with the performance of sports personalities.

6.8 CONCLUSION

This chapter analyses the importance of elastane fibres and their role in different sportswear apparel. The chapter summarizes the effect of the inclusion of elastane yarn in the fabric through different structures namely, woven and knitted fabrics. Further, the processing parameters like elastane percentage, type of yarn, elastane yarn linear density, and its influence on the fabrics' and their garments' overall comfort were outlined. Further, the application of elastane fabric in active sportswear is evaluated and the advantages of using elastane yarn in sportswear were also discussed. The literature collection depicts the lack of direct studies in the field. The research work is either performed from the sports side (like performance analysis of compression sportswear) or evaluated through the textile point of view (like the influence of elastane in comfort/physical properties). However, no direct studies were found in the literature by representing the role of elastane content or composition on the sports as well as comfort aspects of the sportswear.

REFERENCES

1. Sportswear Market Size, Share & Trends Analysis Report By Product (Shoes, Clothes), By End User (Men, Women, Kids), By Distribution Channel (Online, Retail), By Region, And Segment Forecasts, 2019 – 2025. Grand View Research. https://www.grandviewresearch.com/industry-analysis/sportswear-market (Accessed on November 2022).
2. Performance Escalating in Sportswear Fabric. (2010). Apparel Resources, https://in.apparelresources.com/business-news/manufacturing/performance-escalating-sportswear-fabric/ (Accessed on 17th November 2022).
3. Shishoo, R. (2015). *Textile for sportswear*. Cambridge: Woodhead Publishing.
4. Mounir Hassan Khadijah Qashqary, Hany A. Hassan, Ebraheem Shady, & Mofeda Alansary. (2012). Influence of sportswear fabric properties on the health and performance of athletes. *Fibers & Textiles in Eastern Europe*, 4(93), 82–88.
5. Hayes, S.G., & Venkatraman, P. (2016). *Materials and technology for sportwear and performance apparel*. Boca Raton, FL: Taylor & Francis Group.
6. Shishoo, R. (2015). *Textile for sportswear*. Cambridge: Woodhead Publishing.
7. Sewport. (2021). What is elastane fabric: Properties, how its made and where, https://sewport.com/fabrics-directory/elastane-fabric (Accessed on November 2022).
8. Eryuruk, S.H., & Kalaoglu, F. (2016). Analysis of the performance properties of knitted fabrics containing elastane. *International Journal of Clothing Science and Technology*, 28(4), 463–479. https://doi.org/10.1108/IJCST-10-2015-0120
9. Haberstock, H. (1990). Special polyester yarn knitted fabric for sportswear with optimum functional and physiological properties. *Milliand International*, 6, E125–E127.
10. Manshahia, M., & Das, A. (2014). High active sportswear – A critical review. *Indian Journal of Fiber& Textile Research*, 39, 441–449.
11. Chakraborty, D.D., (2013). Functional and interactive sportswear. *Asian Textile Journal*, 22(9), 69.
12. Kothari, V.K. (2003). Fibers and fabrics for active sportswear. *Asian Textile Journal*, 12(3), 55–61.

13. Slater, K. (1977). Comfort properties of textiles. *Textile Progress*, 9(4), 12–15.
14. Sule, A.D., Sarkar, R.K., &Bardhan M.K. (2004). Development of sportswear for Indian conditions. *Manmade Textiles in India*, 12, 123–129.
15. Janarthanan, M. (2013). Soya protein fiber for textile applications, http://www.indiantextilejournal.com/articles/FAdetails.asp?id=4970
16. Bhat, G., Chand, S., & Yakopson, S. (2001). Thermal properties of elastic fibers. *Thermochim. Acta*, 367, 161–164.
17. Abu Yousuf, Mohammad Anwarul Azim, KaziSowrov, Mashud Ahmed, Rakib Ul Hasan, H.M., & Md. Abdullah Al Faruque. (2014). Effect of elastane on single jersey knit fabric properties – physical & dimensional properties. *International Journal of Textile Science*, 3(1), 12–16.
18. Schulze, U. (1993). Rechts/Links-Rundstrick-BindungenınDur KombinationmitDorlastan. *Wirkerei und Strickerei Technology*, 5, 456.
19. BayazitMarmarali, A. (2003). Dimensional and physical properties of cotton/spandex single jersey fabrics. *Textile Research Journal*, 73, 11.
20. Marmaral, A., Özdil, N., & Dönmez Kretzschmar, S. (2006). Thermal comfort and elastic knitted fabrics. International Conference CIRAT-2, Monastir-Tunisia.
21. Tasmaci, M. (1996). Effects of spandex yarn on single jersey fabrics. *TekstilKonfek*,6, 422–426.
22. Haji, M.M.A. (2013). Physical and mechanical properties of cotton/ spandex fabrics. *Pakistan Textile Journal*, 62(1), 52.
23. Prakash, C., & Thangamani, K. (2010). Establishing the effect of loop length on dimensional stability of single jersey knitted fabric made from cotton/lycra core spun yarn. *Indian Journal of Science and Technology*, 3(3), 287–289.
24. Gokarneshan, N., & Thangamani, K. (2012). A comparative evaluation of the low stress-mechanical properties cotton/spandex and polyester/spandex blend knits. *International Journal of Applied Engineering and Technology*, 2(3), 23–27.
25. Herath, C.N., Bok, C.K., & Han-Yong, J. (2006). Dimensional stability of cotton – spandex interlock structures under relaxation. *Fibers and Polymers* Journal, 8, 105–110.
26. Chathura, N., & Bok, C. (2008). Dimensional stability of core spun cotton/spandex single jersey fabrics under relaxation. *Textile Research Journal*; 78, 209–216.
27. Selin HanifeEryuruk, & Fatma Kalaoglu. (2016). Analysis of the performance properties of knitted fabrics containing elastane. *International Journal of Clothing Science and Technology*, 28(4), 463–47.
28. AlenkaPavko-Čuden. (2015). Skewness and spirality of knitted structures. *Tekstilec*, 58(2), 108–120.
29. AlenkaPavko-Cuden, Ales Hladnik, & FranciSluga. (2013). Loop length of plain single weft knitted structure with elastane. *Journal of Engineered Fibers and Fabrics*, 8(2), 110–120.
30. Özdil, N. (2008). Stretch and bagging properties of denim fabrics containing different rates of elastane. *Fibers and Textiles in Eastern Europe*, 16, 63–67.
31. Hassan, Y.M.E., EL-Salmawy, A., & Almetwally, A. (2010). Performance of woven fabrics containing spandex. *Indian Textile Journal*, 120, 22–27.
32. Reyhaneh Masaeli, Hossein Hasani, & Mohsen Shanbeh. (2015). Optimizing the physical properties of elastic-woven fabrics using Grey–Taguchi method. *The Journal of The Textile Institute*, 106(8), 814–822. DOI: 10.1080/00405000.2014.946341
33. Nirmala, V., & Thilagavathi, G. (2014). Development of woven stretch fabrics and analysis on handle, stretch, and pressure comfort. *The Journal of The Textile Institute*, 106(3), 242–252.
34. Babaarslan, O. (2001). Method of producing a polyester/viscose core-spun yarn containing spandex using a modified ring spinning frame. *Textile Research Journal*, 71, 367–371.

35. Al-Ansary, M.A.R. (2011). Effect of spandex ratio on the properties of woven fabrics made of cotton/spandex spun yarn. *Journal of American Science*, 7, 63–67.
36. DunjaSajnGorjanc, & MatejkaBizjak. (2014). The influence of constructional parameters on deformability of elastic cotton fabrics. *Journal of Engineered Fibers and Fabrics*, 9(1), 38–46.
37. Choudhary, A.K., Sheena Bansal, & Nikhil Lodha. The effect of twist multiplier, elastane percentage and pick density on denim quality. *Trends in Textile Engineering and Fashion Technology*, 1(4), 103–108.
38. Çelik, H., & Kaynak, H.K. (2017). An investigation on the effect of elastane draw ratio on air permeability of denim bi-stretch denim fabrics. 17th World Textile Conference AUTEX 2017 – Textiles – Shaping the Future, IOP Conference Series: Materials Science and Engineering. https://iopscience.iop.org/article/10.1088/1757-899X/254/8/082007
39. Das, A., & Chakraborty, R. (2013). Studies on elastane-cotton core-spun stretch yarns and fabrics: Fabric low-stress mechanical characteristics. *Indian Journal of Fiber and Textile Research*, 38, 340–348.
40. Qadir, B., Tanveer, H., Mumtaz, M. (2012). Effect of elastane denier and draft ratio of core-spun cotton weft yarns on the mechanical properties of woven fabrics. *Journal of Engineered Fibers & Fabrics,* 7(4), 23–31.
41. Mecheels, J. (1998). *Body – Climate – Clothing: How does our clothing work?*. Berlin: Schiele & Schoenen.
42. Brotherhood, J. R. (2007). Heat stress and strain in exercise and sport. *Journal of Science Medicine Sport*, 11, 6–19.
43. Bartels, V.T. (2005). Physiological comfort of sportswear. In R Shishoo (Ed.), *Textiles in Sport* (pp. 176–203). Cambridge, England: Wood head Publishing in Textiles.
44. Rigby, D. (1995). Sportswear and fashion, *World Sports Active Wear*, 07, 32.
45. Benkö, T., Cooke, E.A., McNally, M.A., & Mollan, R.A. (2001). Graduated compression stockings: Knee length or thigh length. *Clinical Orthopaedics and Related Research*, 383, 197–203.
46. Bochmann, R.P., Seibel, W., Haase, E., Hietschold, V., Rödel, H., & Deussen, A. (2005). External compression increases forearm perfusion. *Journal of Applied Physiology*, 99(6), 2337–2344.
47. Compression Wear and Shapewear Market by Product Type. (2016). Application (Performance & recovery and body shaping & lifestyle), gender (male and female), distribution channel (multi-retail stores, specialty retail stores, and online channels) – Global opportunity analysis and industry forecasts, 2014–2022. https://www.alliedmarketresearch.com/compression-wear-shapewear-market (Accessed on November 2022).
48. Troynikov, O., Ashayeri, E., Burton, M., Subic, A., Alam, F., & Marteau, S. (2010). Factors influencing the effectiveness of compression garments used in sports. *Procedia Engineering*, 2(2), 2823–2829. https://doi.org/10.1016/j.proeng.2010.04.073
49. Ivan Struhár, Michal Kumstát, & Dagmar MockRálová. (2018). Effect of compression garments on physiological responses after uphill running. *Journal of Human Kinetics,* 61, 119–129.
50. Wolfgang Kemmler, Simon Von Stengel, Christina Ko Ckritz, Jerry Mayhew, Alfred Wassermann, & Ju Rgen Zapf. (2009). Effect of compression stockings on running performance in men runners. *Journal of Strength and Conditioning Research*, 23(1), 101–105.
51. de Glanville, K.M., & Hamlin, M.J. (2012). Positive effect of lower body compression garments on subsequent 40-km cycling time trial performance. *Journal of Strength and Conditioning Research,* 26(2), 480–486.

52. YunusTurgayErten, TurkerSahinkaya, EnginDinc, Bekir ErayKilinc, Bulent Bayraktar, & Mehmet Kurtoglu. (2016). The effects of compression garments and electrostimulation on athletes' muscle soreness and recovery. *Journal of Exercise Rehabilitation*, 12(4), 308–313.

53. Ali, A., Caine, M.P., & Snow, B.G. (2007). Graduated compression stockings: Physiological and perceptual responses during and after exercise. *Journal of Sports Sciences*, 25(4), 413–419. https://doi.org/10.1080/02640410600718376

54. Bieuzen, F., Brisswalter, J., Easthope, C., Vercruyssen, F., Bernard, T., & Hausswirth, C. (2014). Effect of wearing compression stockings on recovery after mild exercise induced muscle damage. *International Journal of Sports Physiology and Performance*, 9(2), 256–264. https://doi.org/10.1123/ijspp.2013-0126

55. Rimaud, D., Calmels, P., Roche, F., Mongold, J.-J., Trudeau, F., & Devillard, X. (2007). Effects of graduated compression stockings on cardiovascular and metabolic responses to exercise and exercise recovery in persons with spinal cord injury. *Archives of Physical Medicine and Rehabilitation*, 88(6), 703–709. https://doi.org/10.1016/j.apmr.2007.03.023.10

56. Miyamoto, N., & Kawakami, Y. (2014). Effect of pressure intensity of compression shorttight on fatigue of thigh muscles. *Medicine and Science in Sports and Exercise*, 46(11), 2168–2174. https://doi.org/10.1249/mss.0000000000000330

57. Siddique Hafiz Faisal, Mazari Adnan, Havel Kaantonin, & Hussain Tanveer. (2018). Effect of elastane linear density on compression pressure of V-shaped compression socks. *Industria Textile*, 69, 2. DOI: 10.35530/IT.069.02.1433

58. Bera, M., Chattopadhay, R., & Gupta, D. (2014). The effect of fiber blend on comfort characteristics of elastic knitted fabrics used for pressure garments, *Journal of The Institution of Engineers (India): Series E*, 95(1), 41–47.

59. Barhoumi, H., Marzougui, S., & Ben Abdessalem, S. (2018). Influence of manufacturing parameters of knitted compression fabric on interface pressure. *Indian Journal of Fiber and Textile Research*, 43, 426–433.

60. Diana Ališauskienė, & Daiva Mikučionienė. (2014). Prediction of compression of knitted orthopaedic supports by inlay-yarn properties. *materials science (medžiagotyra)*, 20(3), 311–314.

61. Kandhavadivu, P., & Gopalakrishnan, D. (2021). Localized compression clothing for improved sports performance. In V. Midha & A. Mukhopadhyay (Eds.), *Recent trends in traditional and technical textiles*. Singapore: Springer. https://doi.org/10.1007/978-981-15-9995-8_8

62. Muhammad Maqsood, Tanveer Hussain, Mumtaz Hasan Malik, & Yasir Nawab. (2016). Modeling the effect of elastane linear density, fabric thread density, and weave float on the stretch, recovery, and compression properties of bi-stretch woven fabrics for compression garments. *The Journal of The Textile Institute*, 107(3), 307–315. http://dx.doi.org/10.1080/00405000.2015.1029809

63. Ali, A., Creasy, R.H., & Edge, J.A. (2011). The effect of graduated compression stockings on running performance. *Journal of Strength and Conditioning Research*, 25(5), 1385–1392.

64. Duffield, R., Cannon, J., & King, M. (2010). The effects of compression garments on recovery of muscle performance following high-intensity sprint and plyometric exercise. *Journal of Science and Medicine in Sport*, 13(1), 136–140.

65. Brophy Williams, N., Gliemann, L., Fell, J., Shing, C., Driller, M., Halson, S., & Askew, C.D. (2014). Haemodynamic changes induced by sports compression garments and changes in posture. Proceedings of the 19th Annual congress of the European College of Sport Science; Amsterdam, Netherlands.

66. Hooper, D.R., Dulkis, L.L., Secola, P.J., Holtzum, G., Harper, S.P., Kalkowski, R.J., Comstock, B.A., Szivak, T.K., Flanagan, S.D., Looney, D.P., DuPont, W.H., Maresh, C.M., Volek, J.S., Culley, K.P., & Kraemer, W.J. (2015). Roles of an upper-body compression garment on athletic performances. *Journal of Strength and Conditioning Research*, 29(9), 2655–2660.

67. Tomas Venckunas, Eugenijus Trinkunas, Sigitas Kamandulis, Jonas Poderys, Albinas Grunovas, & Marius Brazaitis. (2014). Effect of lower body compression garments onhemodynamics in response to running session. *The Scientific World Journal*, 2014(353040), 1–10. http://dx.doi.org/10.1155/2014/353040

68. Duffield, R., & Portus, M. (2007). Comparison of three types of full-body compression garments on throwing and repeat-sprint performance in cricket players. *British Journal of Sports Medicine*, 41(7), 409–414.

69. Sperlich, B., Haegele, M., Achtzehn, S., Linville, J., Holmberg, H.-C., & Mester, J. (2010). Different types of compression clothing do not increase sub-maximal and maximal endurance performance in well-trained athletes. *Journal of Sports Sciences*, 28(6), 609–614.

70. Praburaj, V., & David, T. (2016). Applications of compression sportswear. In Book. In H. Steven George & V. Praburaj (Eds.), *Materials and technology for sportswear and performance apparel* (pp 171–204). Boca Raton, FL: CRC Press, Taylor & Francis Group USA.

71. Craik, J. (2011). The fastskin revolution: From human fish to swimmingandroids. *Culture Unbound*, 3, 71–82.

72. Craig Pickering. (2020). The role of compression garments in performance and recovery. https://simplifaster.com/articles/compression-garments-performance-recovery (Accessed on November 2022).

73. Brown, F., Gissane, C., Howatson, G., van Someren, K., Pedlar, C., & Hill, J. (2017). Compression garments and recovery from exercise – A meta-analysis. *Sports Medicine*, 47(11), 2245–2267. https://doi.org/10.1007/s40279-017-0728-9

74. Cesar Augusto da Silva, Lucas Helal, Roberto Pacheco da Silva, KarlyseClaudino Belli1, Daniel Umpierre, & Ricardo Stein. (2018). Association of lower limb compression garments during high- intensity exercise with performance and physiological responses: A systematic review and meta-analysis. *Sports Medicine*, 48, 1859–1873. https://doi.org/10.1007/s40279-018-0927-z

75. Mourad, M.M., Elshakankery, M.H., & Alsaid, A.A. (2012). Physical and stretch properties of woven cotton fabrics containing different rates of spandex. *Journal of American Science*, 8, 567–572.

76. Hatice Kubra Kaynak. (2017). Optimization of stretch and recovery properties of woven stretch fabrics. *Textile Research Journal*, 87(5), 582–592.

77. Selin Hanife, & Eryuruk Fatma Kalaoglu. (2016). Analysis of the performance properties of knitted fabrics containing elastane. *International Journal of Clothing Science and Technology*, 28(4), 463–479. http://dx.doi.org/10.1108/IJCST-10-2015-0120

78. Payal Bansal, Subhankar Maity, & Sujit Kumar Sinha. (2020) Elastic recovery and performance of denim fabric prepared by cotton/lycra core spun yarns. *Journal of Natural Fibers*, 17(8), 1184–1198. https://doi.org/10.1080/15440478.2018.1558151

79. Osman Gökhan Ertaş, Belkıs Zervent Ünal, & Nihat Çelik. (2016). Analyzing the effect of the elastane-containing dual-core weft yarn density on the denim fabric performance properties. *The Journal of The Textile Institute*, 107(1), 116–126. https://doi.org/10.1080/00405000.2015.1016319

80. Senthilkumar, M. (2014). Dynamics of elastic knitted fabrics for tight fit sportswear. Doctoral Thesis, Anna University, Chennai, India. http://hdl.handle.net/10603/15503

81. Senthilkumar, M., & Anbumani, N. (2011). Dynamics of elastic knitted fabrics for sportswear. *Journal of Industrial Textiles*, 41, 13–24.

82. Senthilkumar, M., Anbumani, N., & Mario de Araujo. (2011). Elastic properties of spandex plated cotton knitted fabric. *Journal of the Institution of Engineers (India), Part TX,* 92.1–5.

83. Gazzah, M., & Jaouachi, B. (2014). Evolution of residual bagging height along knitted fabric lengths. *Research Journal of Textile and Apparel,* 18(4), 70–75. http://dx.doi.org/10.1108/RJTA-18-04-2014-B008

84. Amany Khalil, Abdelmonem Fouda, Pavla Těšinová, & Ahmed S. Eldeeb. (2021). Comprehensive assessment of the properties of cotton single jersey knitted fabrics produced from different lycra states. *AUTEX Research Journal,* 21(1), 71–78. http://dx.doi.org/10.2478/aut-2020-0020

85. Manshahia, M., & Das, A. (2014). Thermo-physiological comfort of compression athletic wear. *Indian Journal of Fiber and Textile Research,* 39, 139–146.

Index

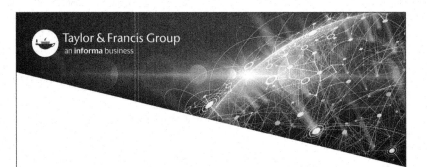

Printed in the United States
by Baker & Taylor Publisher Services